Fundamentals of Electric Circuits

SEVENTH EDITION

LAB MANUAL

DAVID A. BELL

OXFORD
UNIVERSITY PRESS

OXFORD
UNIVERSITY PRESS

70 Wynford Drive, Don Mills, Ontario M3C 1J9
www.oupcanada.com

Oxford University Press is a department of the University of Oxford.
It furthers the University's objective of excellence in research, scholarship,
and education by publishing worldwide in

Oxford New York

Auckland Cape Town Dar es Salaam Hong Kong Karachi
Kuala Lumpur Madrid Melbourne Mexico City Nairobi
New Delhi Shanghai Taipei Toronto

With offices in

Argentina Austria Brazil Chile Czech Republic France Greece
Guatemala Hungary Italy Japan Poland Portugal Singapore
South Korea Switzerland Thailand Turkey Ukraine Vietnam

Oxford is a trade mark of Oxford University Press
in the UK and in certain other countries

Published in Canada
by Oxford University Press

Copyright © Oxford University Press Canada 2009

The moral rights of the author have been asserted

Database right Oxford University Press (maker)

First published 2009

Library and Archives Canada Cataloguing in Publication

Bell, David A., 1930–

Laboratory manual for Fundamentals of electric circuits, seventh edition / David A. Bell.

ISBN 978-0-19-543036-3

1. Electric circuits—Textbooks. 2. Electric circuit analysis—Data processing—Textbooks. I. Title.

TK454.B44 2009 Suppl.   621.319'2   C2009-903675-4

Cover image: NASA

Oxford University Press is committed to our environment. This book is printed on Forest
Stewardship Council® certified paper, harvested from a responsibly managed forest.

Printed in Canada

6 7 8 - 22 21 20

CONTENTS

PREFACE

This laboratory manual has been prepared for use with my book *Fundamentals of Electric Circuits*, 7th Edition. The manual commences with a discussion of rules to be followed in the laboratory, and methods of report writing. Procedures are then given for performing twenty-seven laboratory investigations designed to support the textbook material.

Every investigation has:

- a title
- an introduction that briefly describes the investigation
- a list of required equipment and components
- circuit diagrams and connection diagrams
- a step-by-step procedure to be followed
- a laboratory record sheet for recording data
- an analysis section for processing the data

Each investigation can normally be completed within a two hour period. The procedures contain references to the textbook; however, the circuit diagrams and connection diagrams allow the investigations to be performed without using the book.

David Bell

LABORATORY INVESTIGATIONS AND REPORTS

Reasons for Laboratory Investigations

Laboratory investigations are an important part of your technical education. A well-performed investigation will demonstrate a circuit, or a theory that you are studying, and thus improve your knowledge of the subject. In the laboratory you will become familiar with equipment that will be encountered again in employment. You will also have opportunities to gain skills in recording and analyzing experimental data, and in preparing technical reports.

In the Laboratory

You will work alone or in groups at the discretion of your instructor.

You will follow the procedure listed in this manual for each experiment; however, it can be helpful also have your text book with you in the laboratory.

Most of the equipment that you use in the laboratory is quite expensive. Use it carefully to avoid damage.

Remember that *ELECTRICITY CAN BE LETHAL*; so take care not to risk the safety of yourself or others. If you have any doubt about matters of safety, consult your instructor.

Follow the procedure exactly as listed for each experiment. The procedure is designed to demonstrate the theory under investigation as efficiently as possible. Failure to follow the procedure exactly may result in incorrect data.

Record the experimental data on the record sheet provided, or in an approved laboratory notebook, as required by your instructor.

At the beginning of each investigation, write down the title of each experiment, the date, and the names of any partner working with you. The experimental data should be recorded as neatly as possible. Where procedure is repeated because of error, cross out but do not erase the faulty data.

When the investigation is completed, switch off all equipment and disconnect the supply cords. Return all instruments, cables, components, etc., to their normal storage locations. Instruments should have their supply cords coiled, and dust covers (if available) should be placed over them. Leave your work area clean.

Do not waste any time in the laboratory. Your investigations must be completed, and your work area cleared by the end of the assigned time period. Another class will normally be taking over at the end of your laboratory period. If you do not complete the investigations in the time allocated, you may be required to finish the experiment at another time.

The Laboratory Report

As well as being competent in the use of equipment, a good technician, technologist, or engineer must be able to communicate effectively. For this reason he/she must become proficient in technical writing.

You will be required to prepare a report on each experiment you perform. This report may be written on numbered pages stapled together, or in an approved laboratory notebook, as required by your instructor.

Use good English in your report, and set the work out neatly. Try to write concisely and clearly. Do not write a paragraph where a sentence can contain all the information you wish to convey. You may find it possible to submit some of your technical reports to your English instructor to obtain a grading in that subject.

You are required to report only the experimental results and your analysis of the results. Unless otherwise instructed, do not copy the procedure from the laboratory manual.

Laboratory reports are to be entirely your own work. You may not co-operate with another student in their preparation.

Each report should contain the following:

The *Name of the Investigation*, and the *Date* performed.

Your *Name*, and the names of any *Laboratory Partners*.

Data in the form of results obtained for each step in the procedure. This should follow the order of procedure and use the numbering system listed in the laboratory manual.

Analysis of the experimental data. This will include graphs, calculations, etc. Attempts should be made to explain the results obtained and to answer all questions listed. Any deviations from expected results should be discussed.

Conclusions, in which you should attempt to define the theory that has been demonstrated or the lessons taught by the experiment.

LABORATORY INVESTIGATION 1
Using Digital and Analog Multimeters

Introduction

A digital multimeter is first connected to measure the output voltage of a dc power supply. The power supply voltage is adjusted, and the measured quantities are recorded. Another digital meter is connected to monitor the current flowing from the power supply to a 6 V lamp. Once again, the power supply voltage is adjusted, and the measured quantities are recorded. The digital instruments are also used as ohmmeters to measure the resistance of a decade resistance box. The process is repeated using analog instruments to measure voltage, current, and resistance.

Equipment

DC Power Supply—(0–30 V, 250 mA)
Two Digital Multi-function Meters
Two Analog Multi-function Meters
6 V, 1 W Filament Lamp
Decade Resistance Box—(0–10 kΩ)

Procedure 1   Digital Voltmeter

1-1   Check that the power supply is switched *off*, and that its voltage control is set for zero output.

1-2   Connect the digital multimeter to the power supply as illustrated in Fig. L1-1, taking care to correctly set the function selector for voltage measurement and to use the voltage terminals.

1-3   Switch *on* the power supply and adjust its output to 6 V, as indicated on the power supply meter.

1-4   Observe the indication on the voltmeter, and record the measure voltage on the Laboratory Record Sheet.

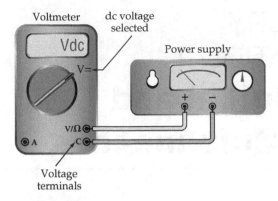

Figure L1-1   Digital multimeter used as a voltmeter.

1-5   Adjust the power supply voltage successively through 5 V, 4 V, 3 V, 2 V, and 1 V, as indicated on the power supply meter. For each voltage level, record the measure voltage on the Laboratory Record Sheet.

1-6   Return the power supply output to 6 V, and then briefly reverse the connections to the voltmeter, and note the effect.

Procedure 2   Digital Ammeter

2-1   Check that the power supply is switched *off*, and that its voltage control is set for zero output.

2-2   Connect the circuit components as illustrated in Fig. L1-2, taking care to correctly set the function selector on each instrument and to use the correct terminals.

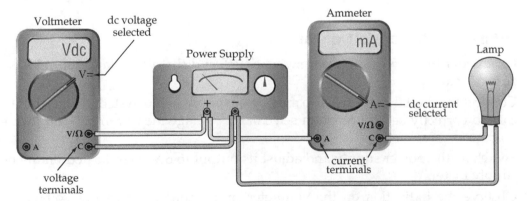

Figure L1-2   Digital voltmeter and ammeter connected to measure voltage and current supplied to a lamp.

2-3   Check that the instrument used as a voltmeter has its function selector set for measuring dc voltage, and that the connections are made to the voltage terminals. (*Wrong function selection and wrong terminal connections are common errors.*)

2-4   Check that the instrument used as an ammeter has its function selector set for measuring dc mA, and that the connections are made to the current terminals.

2-5   Switch *on* the power supply, and adjust its output to 6 V, as indicated on the voltmeter.

2-6   Observe the indication on the ammeter, and record the measured current on the Laboratory Record Sheet.

2-7   Adjust the power supply voltage successively through 5 V, 4 V, 3 V, 2 V, and 1 V as measured by the voltmeter, and record the measured current for each voltage level.

2-8   Return the power supply output to 6 V, then briefly reverse the connections to the ammeter, and note the effect.

Procedure 3   Digital Multimeter as an Ohmmeter

3-1   Set the multimeter function switch for measuring resistance, and then connect it to measure the resistance of the decade box as illustrated in Fig. L1-3.

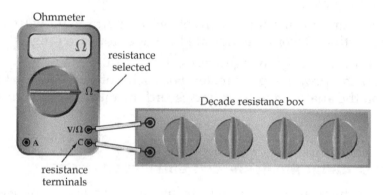

Figure L1-3   Digital multimeter used for measuring resistance.

3-2   Use the instrument to measure the resistance of the decade box at several settings over the entire range of the box. Record the resistance settings of the box and the corresponding measured resistances.

Procedure 4   Analog Multimeter as a DC Voltmeter

4-1   Set the Analog Multimeter to function as a dc voltmeter on its highest range. Check its mechanical zero, and adjust as necessary.

4-2   Check that the power supply is switched *off*, and that its voltage control is set for zero output.

4-3   Connect the voltmeter across the dc power supply terminals; *plus* terminal to *plus* terminal, and *minus* to *minus,* as in Fig. L1-4.

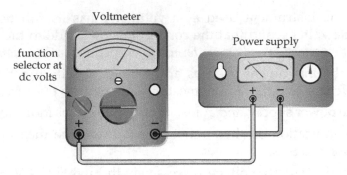

Figure L1-4 Analog Multimeter used for measuring dc voltage.

4-4 Switch *on* the power supply and adjust its output to 5 V as indicated on the power supply meter. If the analog voltmeter deflection is near zero, switch the voltmeter to progressively lower ranges until a *small* readable deflection is obtained.

4-5 Reverse the connections to the voltmeter, note the effect and return the connections to the correct polarity.

4-6 Switch the voltmeter to its next lower voltage range. Carefully observe and record the indicated voltage.

4-7 Repeat Procedure 4-6 through each voltage range to the lowest range that gives on-scale deflection. Record the indicated voltages for each range.

4-8 Adjust the power supply voltage successively through 5 V, 4 V, 3 V, 2 V, and 1 V, as indicated on the power supply meter. For each voltage level, observe the voltage indicated on the analog voltmeter, and record the measured quantities on the Laboratory Record Sheet.

Procedure 5   Analog Ammeter and Voltmeter

5-1 Check the mechanical zero of both instruments, and adjust as necessary.

5-2 Check that the power supply is switched *off*, and that its voltage control is set for zero output.

5-3 Connect the circuit components as illustrated in Fig. L1-5, taking care to correctly set the function selector on each instrument and to use the correct terminals. (*Wrong function selection and wrong terminal connections are common errors.*)

5-4 Set the instrument used as an ammeter to its 200 mA (or next higher) range, and the instrument used as a voltmeter to its 10 V (or next higher) range.

5-5 Switch *on* the power supply and adjust its output to 6 V, as indicated on the analog voltmeter.

5-6 Observe the indication on the ammeter, and record the measure current on the Laboratory Record Sheet. If the pointer deflection is near zero, switch the ammeter to progressively lower ranges until a readable deflection is obtained.

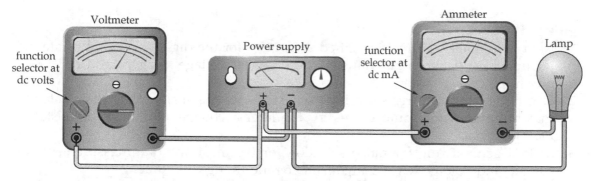

Figure L1-5 Analog multimeters connected for measuring voltage and current.

5-7   Adjust the power supply voltage successively through 5 V, 4 V, 3 V, 2 V, and 1 V, and record the measured current for each voltage level.

5-8   For a brief time interchange the connecting cables at the ammeter terminals. Note that the pointer deflects to the left of zero. Return the connections to the correct polarity.

Procedure 6   Analog Multimeter as an Ohmmeter

6-1   Set the multimeter function switch for measuring resistance, and then zero the instrument. [Short-circuit the terminals, and adjust the ZERO OHMS control for zero indication (right side of the scale) on the *ohms scale*.]

6-2   Use the ohmmeter to measure the resistance of the decade box at several settings over the entire range of the box. Each time the ohmmeter range is changed, the zero should be checked. Record the box resistance values and the corresponding measured resistances.

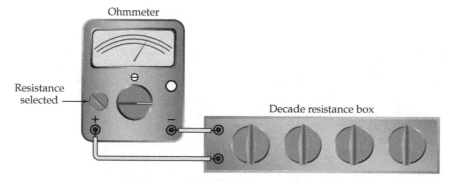

Figure L1-6 Analog multimeter used for measuring resistance.

Analysis

1 Using the manufacturer's specified accuracies for the digital meter, calculate the maximum and minimum values for the largest voltage, current, and resistance measurements.

2 Using the manufacturer's specified accuracies for the analog meter, calculate the maximum and minimum values for the largest voltage, current, and resistance measurements.

3 State a general rule for range selection for the most accurate measurements of current and voltage when using an analog multirange instrument.

4 Why is it necessary to check the mechanical zero of an analog instrument before using it for measurements?

5 Why is it necessary to electrically zero an analog ohmmeter before using it for measurements?

6 State a general rule for range selection for the most accurate measurements of resistance when using an analog multirange instrument.

7 Compare the measurement accuracies of the digital and analog instruments.

Record Sheet L1-1

Record Sheet							Date_____	

Lab. # 1 Digital and Analog Multimeters

Procedure 1–4 & 1–5

Digital Voltmeter

Supply voltage	6	5	4	3	2	1	(V)
Measured voltage							(V)

Procedure 1–6 Reverse connection effect _____

Procedure 2–6 & 2–7

Digital Ammeter

Supply voltage	6	5	4	3	2	1	(V)
Measured current							(mA)

Procedure 2–8 Reverse connection effect _____

Procedure 3

Digital Ohmmeter

Resistance box	10 k	5 k	1 k	100	10	(Ω)
Measured resistance						(Ω)

Procedure 4

Analog Voltmeter

Procedure 4–5 Reverse connection effect _____

Procedure 4–6 & 4–7

Voltmeter range					(V)
Measured voltage					(V)

Procedure 4–8

Supply voltage	6	5	4	3	2	1	(V)
Measured voltage							(V)

Record Sheet L1-2

Record Sheet
Lab. # 1

Procedure 5 Analog Ammeter
Procedure 5–6 & 5–7

Supply voltage	6	5	4	3	2	1	(V)
Measured current							(mA)

Reverse connection effect _____

Procedure 6 Analog Ohmmeter

Ohmmeter range						
Resistance box					(Ω)	
Measured resistance					(Ω)	

LABORATORY
INVESTIGATION 2
Ohm's Law

Introduction

A dc power supply is connected to a decade resistance box, and a voltmeter and ammeter are connected to monitor the resistor voltage and supply current. The voltage, resistance, and current are set to various levels, and each quantity is measured to confirm Ohm's law.

Equipment

DC Power Supply—(0 to 25 V, 100 mA)
DC Ammeter—(100 mA)
Electronic Voltmeter—(30 V).
Decade Resistance Box—(0 to 10 kΩ, 100 mA)

Procedure

1 Check that the power supply is switched *off* and that its output is set for zero voltage.
2 Set the resistance box to 100 Ω and connect the equipment as shown in Fig. L2-1.
3 Switch *on* the power supply and adjust it to give $E = 10$ V.
4 Record the current level on the Record Sheet included with this investigation.
5 Alter R to 250 Ω and adjust the voltage until the ammeter indicates $I = 100$ mA.
6 Record the new level of E.
7 Adjust the power supply to give $E = 25$ V, and alter the decade box until $I = 30$ mA
8 Record the calculated and measured values of R.
9 Adjust the decade box to double the value of R. Record the corresponding values of R, E, and I. Note how the current is affected when R is doubled.
10 Adjust the decade box until R is half the resistance used in Procedure 8. Record the corresponding values of R, E, and I. Note how the current is affected when R is halved.
11 Reset R to 200 Ω and E to 10 V. Record R, E, and I.
12 Adjust the power supply to double the value E to 20 V. Record the new values of R, E, and I, and note how I is affected when E is doubled.

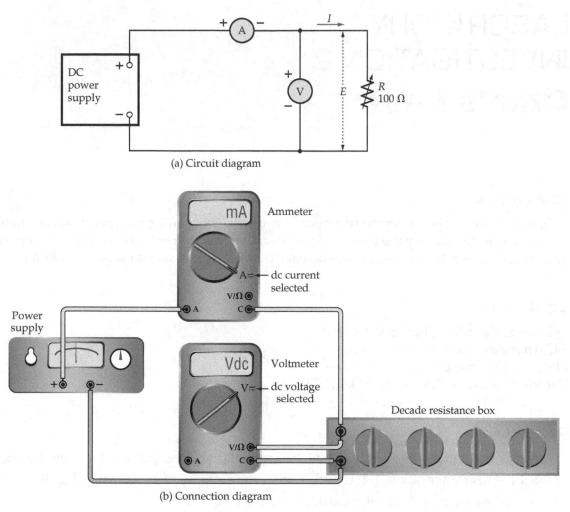

(a) Circuit diagram

(b) Connection diagram

Figure L2-1   Circuit diagram and connection diagram for Ohm's law investigation.

13   Adjust the power supply to give $E = 5$ V (half the voltage level used in procedure 11). Record the level of I, and note how I is affected when E is halved.

14   Adjust the power supply to give $E = 0$ V, and reset R to 250 Ω.

15   Adjust the power supply output to increase E in 5 V steps from 0 V to 25 V. At each step record the levels of E and I indicated on the instruments.

16   Switch *off* the power supply and dismantle the circuit.

Analysis-1 (Ohm's Law)

1-1 Discuss the results of Procedures 3 through 8, as compared to Example 3-3 in the text book.

1-2 Use Ohm's law to explain the effect of doubling the resistance and halving the resistance in a circuit with a constant supply voltage, as demonstrated by the results of procedures 9 through 10.

1-3 Using Ohm's law, explain the effects of doubling and halving the supply voltage to a constant resistance, as demonstrated by the results of Procedures 11 through 13.

1-4 From the results of Procedure 15 plot a graph of current versus voltage for $R = 250 \ \Omega$. Explain the shape of the graph.

Analysis-2 (Power dissipation)

2-1 From the results of Procedures 3 through 6, calculate the power dissipated in R in each case.

2-2 For procedures 7 through 10 calculated the power dissipated in R in each case. Discuss the effects upon power when R is doubled and halved.

2-3 For Procedures 11 through 13 calculate the power dissipated in R in each case. Discuss the effects upon power when E is doubled and halved, and when I is doubled and halved.

Record Sheet L2

Record Sheet					Date	
Lab. #2		Ohm's Law				

Procedure 4
($E = 10$ V, $R = 100$ Ω)

	Measured current	
	Calculated current	

Procedure 6
($I = 100$ mA, $R = 250$ Ω)

	Measured voltage	
	Calculated voltage	

Procedure 8
($E = 25$ V, $I = 30$ mA)

	Measured resistance	
	Calculated resistance	

Procedure 9
(R doubled)

R	E	I	effect on current

Procedure 10
(R halved)

R	E	I	effect on current

Procedure 11
($E = 10$ V, $R = 200$ Ω)

R	E	I

Procedure 12
(E doubled)

R	E	I	effect on current

Procedure 13
(E halved)

R	E	I	effect on current

Procedure 15
($R = 250$ Ω)

E(V)	5	10	15	20	25
I (mA)					

LABORATORY INVESTIGATION 3
Series Resistive Circuit

Introduction

Three measured resistors are connected in series, their total resistance is determined, and the circuit is then connected to a power supply. The voltage drop across each resistor is investigated, and the current flowing is monitored at several points in the circuit. Short-circuits and open-circuits are created to study the effects. A 1.5 V battery is connected alternatively series-aiding and series-opposing with the power supply to observe its effect upon the circuit current. Finally, a two-resistor voltage divider and a potentiometer are investigated.

Equipment

DC Power Supply—(9 V, 50 mA)
DC Ammeter
DC Voltmeter
Ohmmeter
Resistors: $R_1 = 2.2$ kΩ, $R_2 = 1.5$ kΩ, $R_3 = 470$ Ω
Potentiometer: 5 kΩ
Voltage Cell: 1.5 V
Circuit Board

Procedure 1   Resistors in Series

1-1   Using the ohmmeter, carefully measure the resistance of the three resistors: R_1, R_2, and R_3. Record the measured value of each component along with the colour-coded value.

1-2   Connect the resistors in series as shown in Fig. L3-1.

1-3   Use the ohmmeter to measure the total resistance. Record the measured total resistance and the total resistance as determined from the colour code.

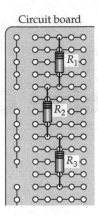

Figure L3-1 Three resistors connected in series.

1-4 Connect the power supply, voltmeter, ammeter, and resistors as in Fig. L3-2.

1-5 Adjust the power supply to give $E = 9$ V.

1-6 Use the voltmeter to measure the voltages V_1, V_2, and V_3 [see Fig. L3-3(a)]. Record each measured voltage.

1-7 Successively connect the ammeter directly in series with each resistor in turn [as illustrated in Fig. L3-3(b)]. For each ammeter position, carefully observe and record the measured current level.

1-8 With the ammeter connected to monitor the power supply current, temporarily short-circuit R_3. Observe and record the new level of current indicated on the ammeter.

1-9 Open-circuit the connection between R_2 and R_3, and connect the voltmeter across the open-circuit. Record the measured voltage level. Also, observe and record the new current level indicated on the ammeter.

1-10 Remove the voltmeter and reconnect the components as shown in Fig. L3-2.

1-11 Connect a 1.5 V voltage cell in series with the power supply and resistors; first in series-aiding, then in series-opposing (see Fig. L3-4). In each case, carefully note the indicated current level.

Procedure 2   Voltage Dividers

2-1 Connect resistors R_1 and R_2 as a voltage divider, as shown in Fig. L3-5. Again use the power supply to provide $E = 9$ V.

2-2 Measure and record the levels of voltages V_1 and V_2.

2-3 Using the ohmmeter, measure the resistance of the 5 kΩ potentiometer between its two outer terminals. Record the measured resistance.

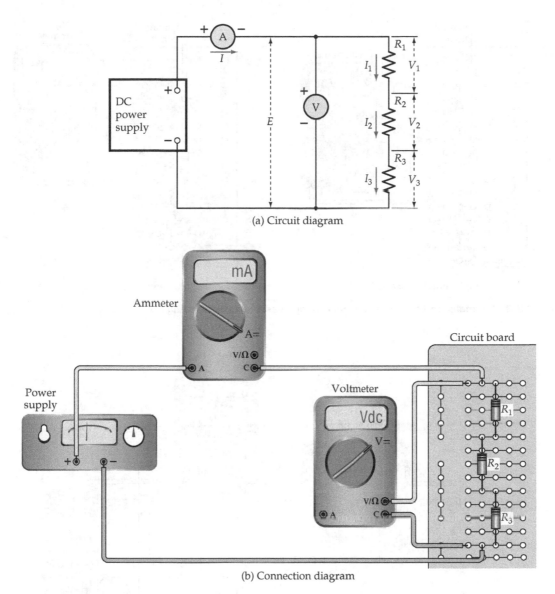

(a) Circuit diagram

(b) Connection diagram

Figure L3-2 Circuit diagram and connection diagram for series resistor circuit.

2-4 Measure the resistance from the centre (moving contact) terminal of the potentiome-ter to one of the outer terminals. Adjust the potentiometer to its extreme clockwise position, and then to its extreme counter-clockwise position. Record the measured resistance values in each case.

2-5 Connect the potentiometer to the power supply as illustrated in Fig. L3-6(a). Set $E = 9$ V and connect the voltmeter to monitor V_o.

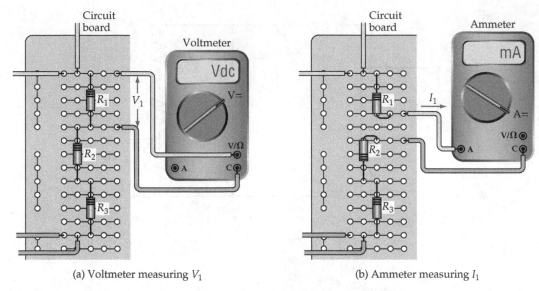

(a) Voltmeter measuring V_1 (b) Ammeter measuring I_1

Figure L3-3   Voltmeter and ammeter connections for measuring resistor voltage and current.

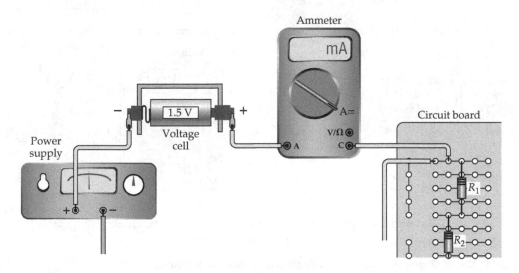

Figure L3-4   Power supply and voltage cell connected series-aiding.

2-6   Carefully adjust the potentiometer to its extreme clockwise position, then to its extreme counter-clockwise position. Observe and record the maximum and minimum values of V_o.

2-7   Connect the resistor R_1 in series with the potentiometer [as in Fig. L3-6(b)], and again apply $E = 9$ V. Repeat Procedure 2-6.

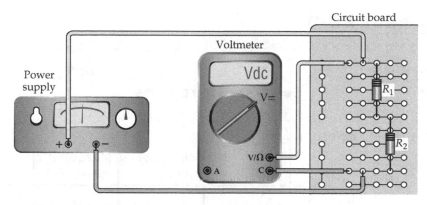

Figure L3-5 Voltage divider circuit.

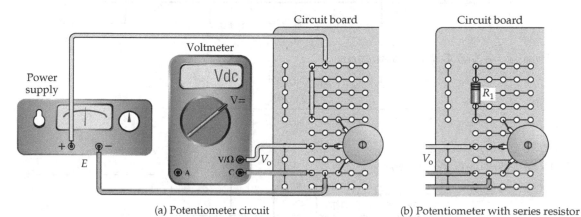

(a) Potentiometer circuit   (b) Potentiometer with series resistor

Figure L3-6 Potentiometer investigation.

Analysis

1   Analyze the three-resistor series circuit to determine total resistance, circuit current, resistor voltages, power dissipated in each resistor, and total circuit power. Record the calculated quantities together with the measured quantities.

2   Calculate the value of circuit current that flows when resistor R_3 is short-circuited. Record this calculated value together with the current level measured in procedure 1-8. Briefly explain.

3   Discuss the voltage and current levels measured for the open-circuit condition in Procedure 1-9.

4   Determine the levels of current that should flow when the 1.5 V cell is connected in series-aiding and series-opposing, as in Procedure 1-11. Also, determine the voltage drop across each resistor for each of the two cases. Relate the voltage drop in each case to Kirchhoff's voltage law.

5   Calculate the voltage levels V_1 and V_2 for the voltage divider and potentiometer circuits investigated in Procedure 2. Compare the calculated and measured voltages. Calculate the power dissipated in each component.

Record Sheet L3-1

Record Sheet	Date
Lab. # 3	Series Resistive Circuits

Procedure 1–1 Resistor values

	Colours	Colour-coded resistance	Measured resistance
R_1			
R_2			
R_3			

Procedure 1–3 Total resistance

	Measured resistance	Calculated resistance
$R_1 + R_2 + R_3$		

Procedure 1–6 Voltages

	E	V_1	V_2	V_3
Measured voltage	9 V			
Calculated voltage	9 V			

Procedure 1–7 Current levels

I_1	I_2	I_3

Procedure 1–8 R_3 short-circuited

Procedure 1–9 R_2R_3 open-circuited

Open-circuit voltage	I

Procedure 1–11 Voltage cell connected

	Series-aiding	Series-opposing
I		

18

Record Sheet L3-2

Record Sheet
Lab. # 3

Procedure 2–2 Voltage divider

V_1	V_2

Procedure 2–3 & 2–4 Potentiometer

Total resistance	Resistance range

Procedure 2–6 & 2–7 Potentiometer voltage

	$V_{(min)}$	$V_{(max)}$
Without R_1		
With R_1		

LABORATORY INVESTIGATION 4
Parallel Resistive Circuits

Introduction

Four measured resistors are connected in parallel, and the parallel circuit resistance is determined. The four (parallel-connected) resistors are then connected to the terminals of a power supply. The terminal voltage of each resistor, the supply current, and the current flowing in each resistor are all measured. An open-circuit is created to observe its effect upon the supply current. Finally a two-resistor current divider circuit is investigated.

Equipment

DC Power Supply—(0 to 24 V, 50 mA)
DC Ammeter
DC Voltmeter
Ohmmeter
Resistors: $R_1 = 2.2$ kΩ, $R_2 = 5.6$ kΩ, $R_3 = 3.3$ kΩ, $R_4 = 4.7$ kΩ
Circuit Board

Procedure 1   Resistors in Parallel

1-1 Use the ohmmeter to measure the resistance R_1, R_2, R_3, and R_4. Record the measured value along with the colour-coded value of each component.

1-2 Connect the four resistors in parallel as shown in Fig. L4-1. *Do not connect any battery or power supply at this time.*

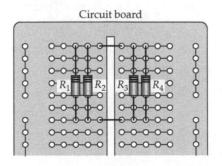

Figure L4-1   Four resistors connected in parallel.

1-3 Use the ohmmeter to measure the resistance of the four resistors in parallel. Record the measured resistance, and calculate the parallel equivalent resistance using the colour-coded values.

1-4 Connect the instruments and resistors as shown in Fig. L4-2.

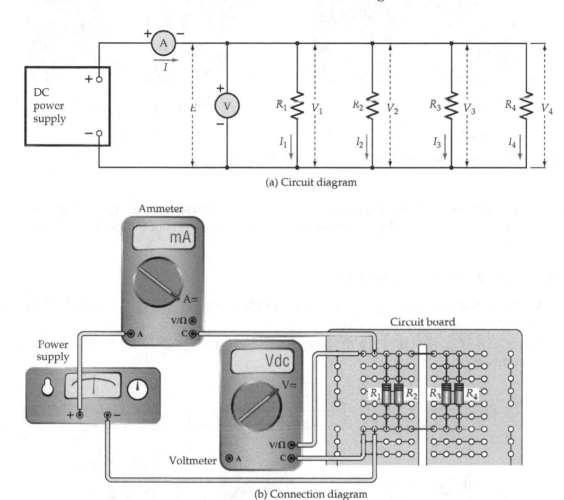

(a) Circuit diagram

(b) Connection diagram

Figure L4-2 Circuit diagram and connection diagram for parallel-connected resistors.

1-5 Adjust the power supply to give $E = 24$ V.

1-6 Use the voltmeter to measure the terminal voltage of each resistor [see Fig. L4-3(a)]. Record each measured voltage.

1-7 Record the level of supply current indicated by the ammeter, and then successively connect the ammeter in series with each resistor in turn to measure I_1, I_2, I_3, and I_4 [see Fig. L4-3(b)]. Record all current levels.

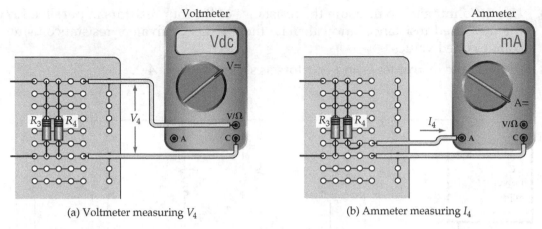

(a) Voltmeter measuring V_4 (b) Ammeter measuring I_4

Figure L4-3 Voltmeter and ammeter connections for measuring V_4 and I_4.

1-8   Reconnect the ammeter to measure the supply current once again. Open-circuit resistor R_1, and carefully observe and record the new level of supply current.

Procedure 2   Current Dividers

2-1   Connect the resistors R_1 and R_2 in parallel as shown in Fig. L4-4 and adjust the power supply to give $E = 9$ V.

2-2   Successively connect the ammeter to measure I, I_1, and I_2 in turn. Record each current level.

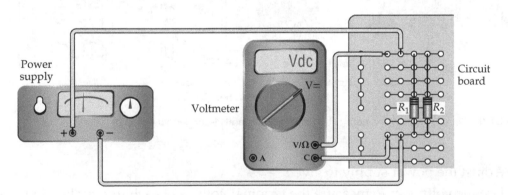

Figure L4-4 Current divider circuit.

Analysis

1 Analyze the four-resistor circuit to determine the parallel equivalent resistance, each resistor current, the power dissipation in each resistor, and the total circuit power. Record the measured and calculated quantities. Relate the measured current levels to Kirchhoff's current law.

2 Convert each resistor to a conductance, and then repeat the analysis of the four-resistor circuit to determine all current levels.

3 Calculate the level of supply current that flows when R_1 is open-circuited. Record the measured and calculated current levels.

4 Analyze the two-resistor parallel circuit to determine the total supply current, then use the current divider equation to calculate each resistor current. Tabulate the calculated and measured current levels.

Record Sheet L4

Record Sheet					Date	
Lab. # 4		Parallel Resistive Circuits				

Procedure 1–1 Resistor values

	Colours	Colour-coded resistance	Measured resistance
R_1			
R_2			
R_3			
R_4			

Procedure 1–3 Total parallel resistance

	Measured resistance	Calculated resistance
$R_1 \| R_2 \| R_3 \| R_4$		

Procedure 1–6 Measured voltages

E	V_1	V_2	V_3	V_4
24 V				

Procedure 1–7 Measured currents

I	I_1	I_2	I_3	I_4

Procedure 1–8 R_1 open-circuited supply current

$I =$ _____

Procedure 2–2 Current divider

I	I_1	I_2

LABORATORY
INVESTIGATION 5
Series-Parallel Circuits

Introduction

Five resistors are measured and then connected to form a series-parallel circuit. Each resistor's terminal voltage and current level is carefully measured. The supply current is monitored, and the effect of open-circuiting and short-circuiting one resistor is investigated.

Equipment

DC Power Supply—(15 V, 50 mA)
DC Ammeter
DC Voltmeter
Ohmmeter
Resistors: $R_1 = 4.7$ kΩ, $R_2 = 39$ kΩ, $R_3 = 27$ kΩ, $R_4 = 5.6$ kΩ, $R_5 = 22$ kΩ
Circuit Board

Procedure

1   Use the ohmmeter to measure the resistance value of each resistor. Record the measured and colour-coded resistances.
2   Connect the components as shown in Fig. L5-1. *Do not connect any battery or power supply at this time.*

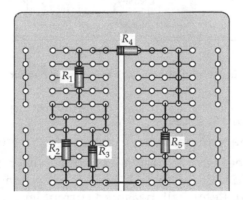

Figure L5-1   Series-parallel resistor circuit.

3 Use the ohmmeter to measure the total circuit resistance. Record the measured value.

4 Connect a power supply, voltmeter, and ammeter to the circuit, as in Fig. L5-2.

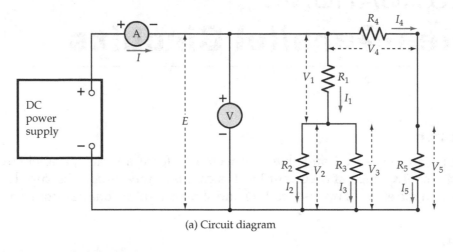

(a) Circuit diagram

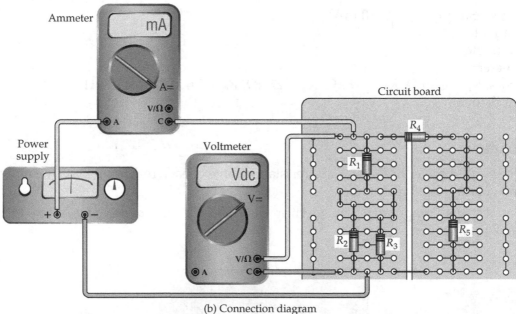

(b) Connection diagram

Figure L5-2 Circuit diagram and connection diagram for series-parallel resistor circuit.

5 Adjust the power supply to give $E = 15$ V.

6 Measure and record the terminal voltage of each resistor [see Fig. L5-3(a)].

7 Record the level of the current drawn from the power supply, then successively connect the ammeter in series with each resistor to determine the currents $I_1, I_2, I_3, I_4,$ and I_5 [see Fig. L5-3(b)]. Record all measured current levels.

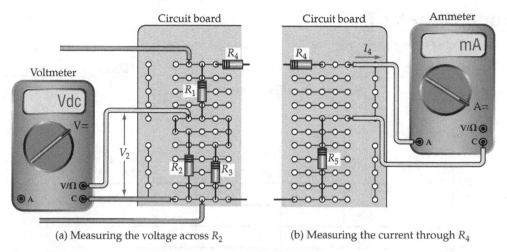

(a) Measuring the voltage across R_2 (b) Measuring the current through R_4

Figure L5-3   Voltmeter and ammeter connections for measuring V_2 and I_4.

8   Once again connect the ammeter to monitor the supply current. Open-circuit resistor R_3 and note the effect on the supply current. Short-circuit R_3 and again note the effect on the supply current.

Analysis

1   Compare the measured values of current, voltage and total circuit resistance to those calculated for Problems 7-4, 7-10, and 7-16 in the text book.

2   Calculate the level of power supply current that flows when R_3 is open-circuited, and when R_3 is short-circuited. Compare the calculated values to the measured current levels.

Record Sheet L5

Record Sheet						Date _____
Lab. # 5		Series-Parallel Resistive Circuits				

Procedure 1 Resistor values

	Colours	Colour-coded resistance	Measured resistance
R_1			
R_2			
R_3			
R_4			
R_5			

Procedure 3 Series-parallel resistance

	Measured resistance	Calculated resistance
R		

Procedure 6 Measured voltages

E	V_1	V_2	V_3	V_4	V_5
15 V					

Procedure 7 Measured currents

I	I_1	I_2	I_3	I_4	I_5

Procedure 8 Open-circuit and short-circuit

	R_3 open	R_3 shorted
Supply current		

LABORATORY INVESTIGATION 6
Resistive Networks

Introduction

Three decade resistance boxes and two power supplies are connected together to form the network shown in Fig. 8-10 in the text book. Voltage and current levels are measured throughout the circuit.

A power supply and decade box are connected to simulate a voltage source, and then to form a current source. In both cases, load resistors are connected to the source output terminals, and the output current and voltage are measured.

Three decade resistance boxes are connected to alternatively form Δ and Y networks. In each case the resistances between pairs of terminals are measured.

Equipment

Two DC Power Supplies—(0 to 30 V, 100 mA)
DC Ammeter
DC Voltmeter
Ohmmeter
Three Decade Resistance Boxes—(0 to 10 kΩ, 20 mA)

Procedure 1   Resistor Network

1-1   Identify the three decade resistance boxes as R_1, R_2, and R_3, and set the resistance values to 120 Ω, 240 Ω, and 200 Ω, as in Fig. 8-10 in the text book.

1-2   Identify the two power supplies as E_1 and E_2, adjust the voltages to 6 V and 12 V respectively, and connect up the circuit as in Fig. L6-1.

1-3   Using a voltmeter carefully check voltages E_1 and E_2 and adjust if necessary.

1-4   Measure voltage V_3 and record it on the record sheet provided.

1-5   Use an ammeter to measure I_1, I_2, and I_3. Record each current level.

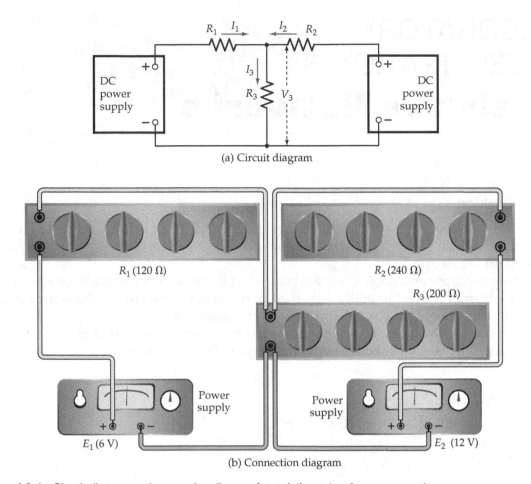

(a) Circuit diagram

(b) Connection diagram

Figure L6-1 Circuit diagram and connection diagram for resistive network measurements.

Procedure 2 Voltage Sources and Current Sources

2-1 Adjust a power supply to give $E = 1$ V and connect a resistor $R_S = 100 \ \Omega$ in series with it to represent a voltage source, as in Fig. L6-2.

2-2 Connect a resistor $R_L = 10$ kΩ across the output terminals of the voltage source. Carefully measure and record V_L and I_L.

2-3 Calculate the values of I_S and R_S for the equivalent current source.

2-4 Connect the current source together with $R_L = 10$ kΩ as in Fig. L6-3. For the current generator use a 30 V power supply with an ammeter and a 2.7 kΩ resistor connected in series with its output terminals. Adjust the power supply voltage to give the calculated source current level.

2-5 Carefully measure and record the levels V_L and I_L.

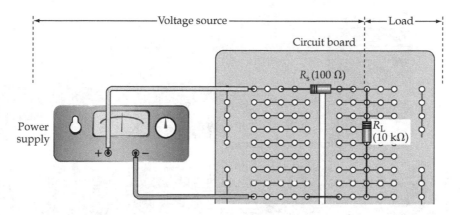

Figure L6-2 Voltage source and load resistor.

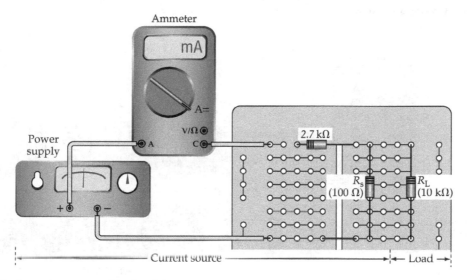

Figure L6-3 Current source and load resistor.

Procedure 3 Δ-Y transformation

3-1 Using three decade resistance boxes set at the appropriate resistance values, construct the delta network in Fig. L6-4, using the resistance valued from Example 8-6 in the text book ($R_{ab} = 500\ \Omega$, $R_{ac} = 400\ \Omega$, $R_{bc} = 300\ \Omega$).

3-2 With an ohmmeter carefully measure the resistances between terminals A and B, A and C, B and C.

3-3 Reconstruct the circuit in the form of the Y network in Fig. L6-5. Set each decade box to the R_a, R_b, and R_c values calculated in Example 8-6.

3-4 Once again measure the resistances at terminals A and B, A and C, B and C.

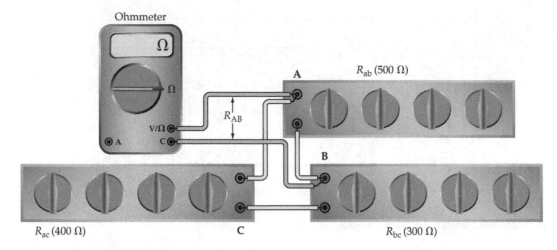

Figure L6-4   Resistor Δ network.

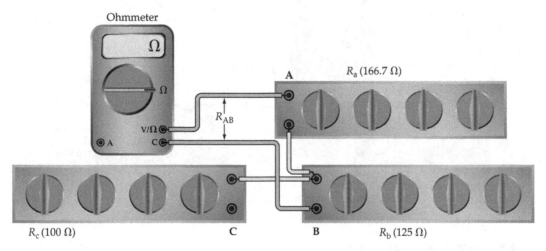

Figure L6-5   Resistor Y network.

Analysis

1   Compare the measured voltage and current levels from Procedure 1 to the values calculated in Example 8-2 in the text book.
2   Compare the load current and load voltage measurements made on the voltage source to those made on the current source, as investigated in Procedure 2.
3   For the results of Procedure 3 compare the terminal resistance values for the Δ and Y networks.

Record Sheet L6

Record Sheet   Date _____

Lab. # 6   Resistive Networks

Procedure 1 Resistor network
Procedure 1–4 and 1–5

	V_3	I_1	I_2	I_3

Procedure 2 Voltage sources and current sources
Procedure 2–2

	V_L	I_L

Procedure 2–3 and 2–5

	I_S	R_S	V_L	I_L

Procedure 3 Δ-networks and Y-networks
Procedure 3–2

	A-B	A-C	B-C
Resistance			

Procedure 3–3

	A-B	A-C	B-C
Resistance			

LABORATORY INVESTIGATION 7
Network Theorems

Introduction

Three network theorems are investigated: Superposition theorem, Thevenin's theorem, and the Maximum Power Transfer theorem. In each case, resistances and power supplies are connected into the appropriate configurations corresponding to circuits shown in the text book. The voltage and current levels throughout each circuit are carefully measured for comparison to the values calculated in the book examples.

Equipment

Two DC Power Supplies—(0 to 30 V, 100 mA)
DC Ammeter
DC Voltmeter
Three Decade Resistance Boxes—(0 to 10 kΩ, 20 mA)
Resistors: 330 Ω, 560 Ω, 680 Ω, 820 Ω

Procedure 1   Superposition Theorem

1-1   Identify the three decade resistance boxes as R_1, R_2, and R_3, and set them to the resistance values shown in Fig. 9-1 in the text book; 120 Ω, 240 Ω, and 200 Ω respectively.

1-2   Identify the two Power Supplies as E_1 and E_2, adjust the voltages to 6 V and 12 V respectively, and connect up the circuit as shown in Fig. L7-1. Note that the ammeter is connected in series with R_3.

1-3   Check the voltage levels of E_1 and E_2 and adjust if necessary, and then carefully observe and record the level of current I_3.

1-4   Disconnect voltage E_2 and replace it with a short circuit [as in Fig. 9-1(b) in the text book]. Observe and record the level of current I_a.

1-5   Remove the short-circuit, and reconnect the voltage E_2. Disconnect voltage E_1, and replace it with a short-circuit [as in Fig. 9-1(c) in the text book]. Observe and record current I_b.

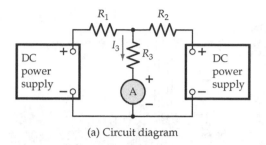

(a) Circuit diagram

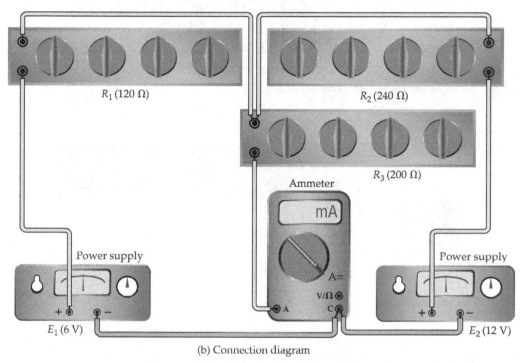

(b) Connection diagram

Figure L7-1 Circuit and connection diagram for network theorems investigation.

Procedure 2   Thevenin's Theorem

2-1   Reconstruct the circuit of Fig. L7-1.

2-2   Connect a voltmeter and a load resistor $R_L = 330\ \Omega$ across R_3, as in Fig. L7-2.

2-3   Measure and record the level of the load voltage V_{L1}.

2-4   Remove the 330 Ω load resistor and successively connect $R_{L2} = 560\ \Omega$, $R_{L3} = 680\ \Omega$ and $R_{L4} = 820\ \Omega$. In each case record the levels of V_{L2}, V_{L3}, and V_{L4}.

2-5   Construct the Thevenin equivalent circuit in Fig. L7-3 using a decade resistance box for $R_S = 57.1\ \Omega$. Carefully adjust E_{th} to 5.71 V.

2-6   Successively connect load resistors $R_{L1} = 330\ \Omega$, $R_{L2} = 560\ \Omega$, $R_{L3} = 680\ \Omega$, and $R_{L4} = 820\ \Omega$. In each case, measure and record V_{L1}, V_{L2}, V_{L3}, and V_{L4}.

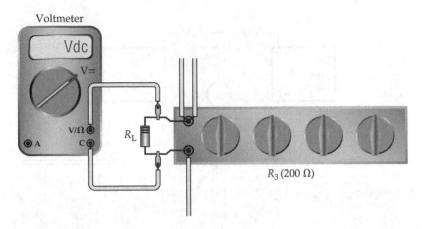

Figure L7-2 Voltmeter and load resistor connected across R_3.

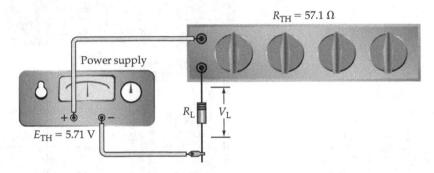

Figure L7-3 Thevenin equivalent circuit with load resistor.

Procedure 3   Maximum Power Transfer Theorem

3-1   Using a dc power supply and two decade resistance boxes, construct a Thevenin equivalent circuit with a variable load, as in Fig. L7-4. Set the voltage to $E_{th} = 10$ V, and the source resistance to $R_{th} = 500$ Ω.

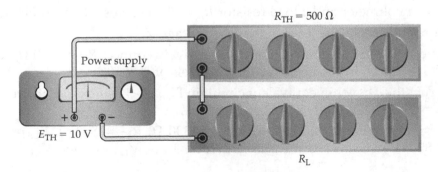

Figure L7-4 Thevenin equivalent circuit with adjustable load resistor.

3-2 Connect a voltmeter to monitor the load voltage V_L.

3-3 Adjust the load resistor through: 50 Ω, 100 Ω, 250 Ω, 500 Ω, 1 kΩ, 2.5 kΩ, and 5 kΩ. For each value of R_L measure and record the level of V_L.

Analysis

1 Compare the levels of I_3, I_a, and I_b measured in Procedure 1 to the calculated values in Example 9-1 in the text book.

2 Compare the voltages measured in Procedure 2-3 and 2-4 to the corresponding voltage values measured in Procedure 2-6. Also use V_{L1} to calculate I_L, and compare it to the value of I_L calculated in Example 9-4 in the text book.

3 From the results of Procedure 3, calculate the values of I_L and P_L for each load resistance. Plot the graphs of V_L, I_L, and P_L approximately to a logarithmic base of R_L (see Fig. 9-10 in the text book).

Record Sheet L7

Record Sheet   Date _____

Lab. # 7   Network Theorems

Procedure 1 Superposition theorem
Procedure 1–3, 1–4, 1–5

	I_3	I_a	I_b

Procedure 2 Thevenin's theorem
Procedure 2–3, and 2–4

R_L	330 Ω	560 Ω	680 Ω	820 Ω
V_L				

Procedure 2–6

R_L	330 Ω	560 Ω	680 Ω	820Ω
V_L				

Procedure 3 Maximum power transfer theorem

$R_L(\Omega)$	50	100	250	500	1 K	2.5 k	5 k
$V_L(V)$							

LABORATORY INVESTIGATION 8
Voltage Cells and Batteries

Introduction

The open-circuit terminal voltage of each of six cells is first measured. The cells are then successively connected in several different arrangements to form a battery. The battery output voltage and internal resistance is investigated for each cell arrangement. Finally, one cell is loaded to supply a constant current, and its terminal voltage is monitored over a period of time.

Equipment

Two Electronic Voltmeters—(0 to 10 V dc)
Two Ammeters—(0 to 100 mA dc)
Two Decade Resistance Boxes—(0 to 500 Ω, 500 mA)
Seven Zinc-Carbon D Cells
Cell Holders

Procedure

1 Identify each cell by number, and use the voltmeter to measure the terminal voltage of each. Record the values as E_1, E_2, etc., on the record sheet provided.

2 Connect four cells in series-aiding as illustrated in Fig. L8-1. Measure the terminal voltage (E) of the battery of four cells.

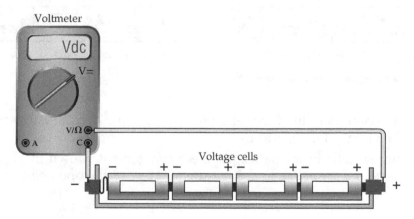

Figure L8-1 Four series-connected cells.

3 Set a decade resistance box to its highest resistance and connect it as a load resistor R_L across the terminals of the four cell battery, as in Fig. L8-2.

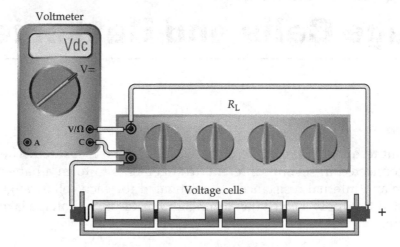

Voltmeter

Figure L8-2 Battery of voltage cells with variable load resistor R_L.

4 Reduce the resistance of R_L until E_o falls by about 0.5 V below the *no-load* terminal voltage E measured in Procedure 2. Record the exact values of E_o and R_L.

5 Disconnect R_L from the battery and reconnect three of the cells series-aiding as shown in Fig. L8-3(a). Measure and record the terminal voltage E.

6 Reconnect one of the cells series-opposing as in Fig. L8-3 (b). Again measure and record the level of E.

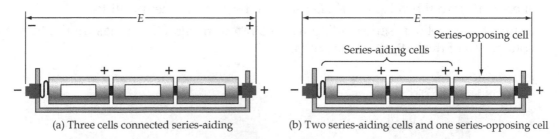

(a) Three cells connected series-aiding   (b) Two series-aiding cells and one series-opposing cell

Figure L8-3 Series-aiding and series-opposing cells.

7 Connect the six cells in the form of a plus-minus supply, as illustrated in Fig. L8-4. Measure the voltages E, E_1, and E_2.

8 Connect the six cells in the series-parallel arrangement shown in Fig. L8-5. *Do not connect the load resistor at this time.* Measure and record the open-circuit voltage E.

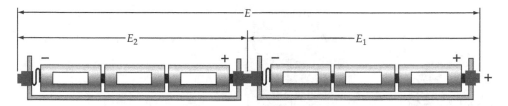

Figure L8-4 Voltage cells connected as a plus-minus supply.

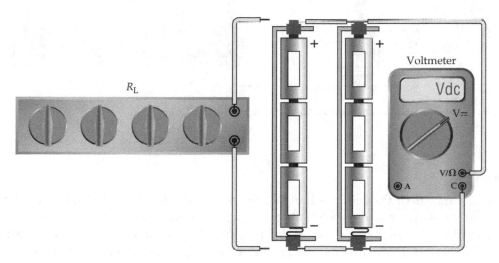

Figure L8-5 Series-parallel battery of cells.

9 Set the decade box used as R_L to its highest resistance, and then connect it across the terminals of the battery.

10 Adjust R_L until the battery terminal voltage E_o falls by approximately 0.5 V from the no-load terminal voltage measured in Procedure 8. Record the exact values of E_o and R_L.

11 Dismantle the battery of cells.

12 Construct the single cell circuit in Fig. L8-6 in which the voltmeter is monitoring the cell terminal voltage, and the ammeter is measuring the load current.

13 Adjust R_L to give $I_L = 100$ mA and record the level of E_o.

14 Leaving the circuit connected, check the level of I_L after a time of 10 minutes. Adjust R_L as necessary to maintain I_L approximately at 100 mA. Record the level of E_o at this time.

15 Repeat Procedure 14 at 10-minute intervals over a period of one hour.

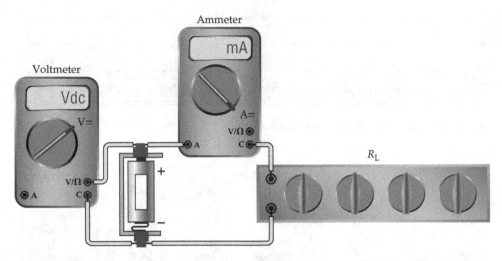

Figure L8-6 Circuit for voltage cell terminal-voltage versus time test.

Analysis

1 For the results of Procedures 1 and 2, check that the terminal voltage of the battery of four cells approximately equals the sum of the individual cell voltages.

2 Using the values of E, E_o, and R_L measured in Procedures 2 and 4, calculate $I = E_o/R_L$ and determine the battery internal resistance r from: $I = E/(r + R_L)$. Also estimate the internal resistance of each cell.

3 Discuss the results of Procedures 5, 6, and 7.

4 From the results of Procedures 8, 9, and 10, calculate the output current and the internal resistance of the six-cell series-parallel battery.

5 Using the values of E_o measured in Procedures 13, 14, and 15, plot a graph of E_o versus time for a constant load current of 100 mA (see Fig. 10-8 in the text book).

Record Sheet L8

Record Sheet   Date _____

Lab. # 8   Voltage Cells and Batteries

Procedure 1 Cell voltages

E_1	E_2	E_3	E_4	E_5	E_6

Procedure 2 Battery no-load voltage $E=$ _____

Procedure 4 Loaded conditions

E_0
R_L

Procedure 5 Three-cell battery $E=$ _____

Procedure 6 One cell series-opposing $E=$ _____

Procedure 7 Plus/minus supply

E_1	E_2	E_3

Procedure 8, 9, 10 Series-parallel battery

E_0	E_0	R_L

Procedure 13, 14, 15 Cell terminal voltage

$t_{(min)}$	0	10	20	30	40	50	60
$E_0(V)$							

LABORATORY INVESTIGATION 9
Magnetic Fields

Introduction

The magnetic force fields surrounding bar magnets are plotted using compasses and iron filings. Compasses are also used to investigate the magnetic fields around current carrying conductors, and at the ends of a solenoid.

Equipment

DC Power Supply—(10 V, 10 A), or 12 V Automobile Battery.
Magnetic Compass—(one or more)
Two Bar Magnets
Iron Filings
Insulated Wire—(about 2 m of 16 gauge or thicker)

Procedure 1   Permanent Magnets

1-1   Place a bar magnet on a bench, then bring a magnetic compass close to one end of the magnet, as illustrated in Fig. 11-1(c) in the textbook. Identify the polarity of the end of the bar magnet and repeat the procedure at the opposite end.

1-2   Repeat Procedure 1-1 with the second bar magnet to identify the polarity of its ends.

1-3   Place a thin sheet of cardboard over one of the bar magnets. Sprinkle iron filings on the cardboard, then tap it gently. The filings should form a pattern similar to that shown in Fig. 11-2(a) in the textbook. Using the record sheet provided, make a sketch of the field pattern; or take a photograph if a camera is available.

1-4   Remove the iron filings and replace the cardboard above the magnet. Use a compass to plot one line-of-force on the cardboard, as illustrated in Fig. 11-2(b) in the book. Repeat the process to obtain plots of several lines of force.

1-5   Using two magnets, a sheet of cardboard, and iron filings, obtain the field pattern shown in Fig. 11-2(c). Repeat the process to obtain plots of several lines of force.

1-6   Repeat Procedure 1-5 to obtain the field patterns illustrated in Fig. 11-4(a) and (b) in the book.

1-7   Place a piece of soft iron, a piece of wood, and a piece of aluminum near a bar magnet. Use a sheet of cardboard and iron filings to investigate the field pattern. The result should be similar to field shown in Fig. 11-3(a) in the textbook.

Procedure 2 Current Carrying Conductors

2-1 Support a piece of cardboard (about 15 cm × 15 cm) in a horizontal position. Vertically pass a piece of insulated cable through a hole in the center of the cardboard, and connect the ends of the cable via an ammeter to the terminals of the 10 A power supply [see Fig. L9-1(a)]. *If an automobile battery is used instead of the power supply include a 1 Ω, 100 W resistor in series with the wire.*

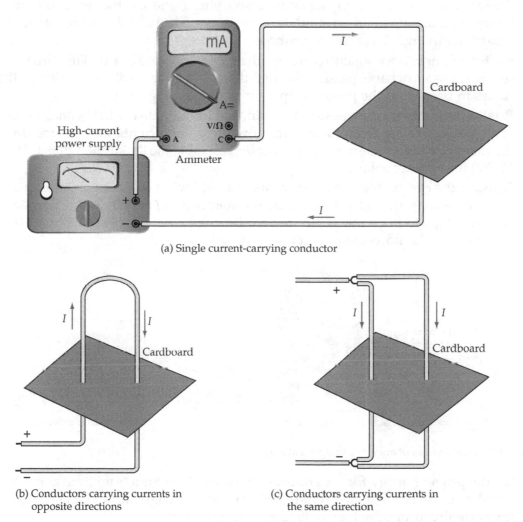

(a) Single current-carrying conductor

(b) Conductors carrying currents in opposite directions

(c) Conductors carrying currents in the same direction

Figure L9-1 Connection arrangements for investigating the magnetic fields around current-carrying conductors.

2-2 Adjust the power supply control to the zero output voltage position, then switch *on* for a *short time* and adjust for a current of approximately 10 A. Use one or more compasses to investigate the direction of the magnetic field around the conductor. See the illustration in Fig. 11-5(b) in the text book.

2-3 Switch *off*, and reverse the terminal connections to reverse direction of the current in the conductor. Switch *on* again for a short time, and once more use compasses to investigate the magnetic field.

2-4 Switch the power supply *off*, and rearrange the cable so that it passes up through one hole in the cardboard, turns around about 10 cm above the cardboard then passes down through another hole about 7 cm from the first hole, as illustrated in Fig. L9-1(b).

2-5 Switch *on* the power supply again for a short time, and use the compasses to investigate the magnetic fields around the conductors. The field should be similar to that illustrated in Fig. 11-7(a) in the textbook.

2-6 Switch *off*, and once again rearrange the cable and cardboard. This time use two separate pieces of cable passing through the holes in the cardboard. Connect the two cables in parallel to the power supply terminals, see Fig. L9-1(c).

2-7 Briefly switch *on* the power supply again, and investigate the fields once more using compasses. Since the conductors are now carrying currents in the same direction, the magnetic field pattern should resemble resemble the illustration in Fig. 11-7(b) or 11-7(c) in the textbook.

2-8 Remove the conductors from the cardboard, and wind one of them around a pencil to create a long thin closely wound coil. Remove the pencil, and if necessary use adhesive tape to keep the coil in shape. With the power supply *off*, connect the ends of the coil to the power supply terminal, (see Fig. L9-2).

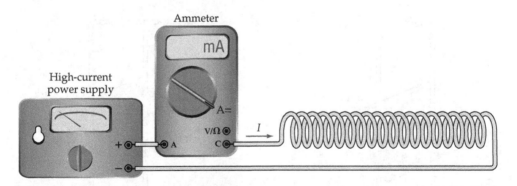

Figure L9-2 Investigation of magnetic field around a coil.

2-9 Set the power supply for zero output, then switch *on* for a brief time and adjust for $I \approx 10$ A. Use a compass to investigate the resultant magnetic field. Fig. 11-8(c) in the textbook illustrates the type of field pattern to expect.

Analysis

1 Explain the shape of each of the field patterns for the permanent magnets, as obtained for Procedure 1.

2 Explain the field patterns obtained for current-carrying conductors and coils in Procedure 2.

Record Sheet L9-1

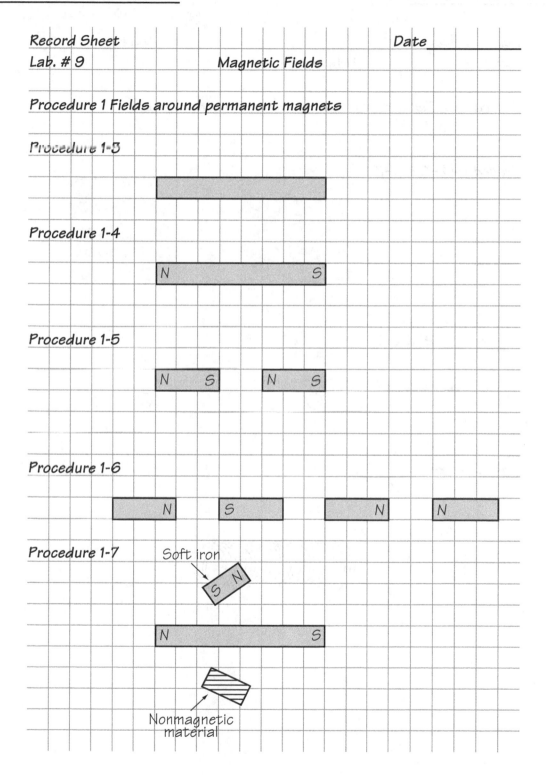

Record Sheet

Lab. # 9   Magnetic Fields

Date_____

Procedure 1 Fields around permanent magnets

Procedure 1-3

Procedure 1-4

N   S

Procedure 1-5

N   S   N   S

Procedure 1-6

N   S   N   N

Procedure 1-7

Soft iron

S N

N   S

Nonmagnetic
material

Record Sheet L9-2

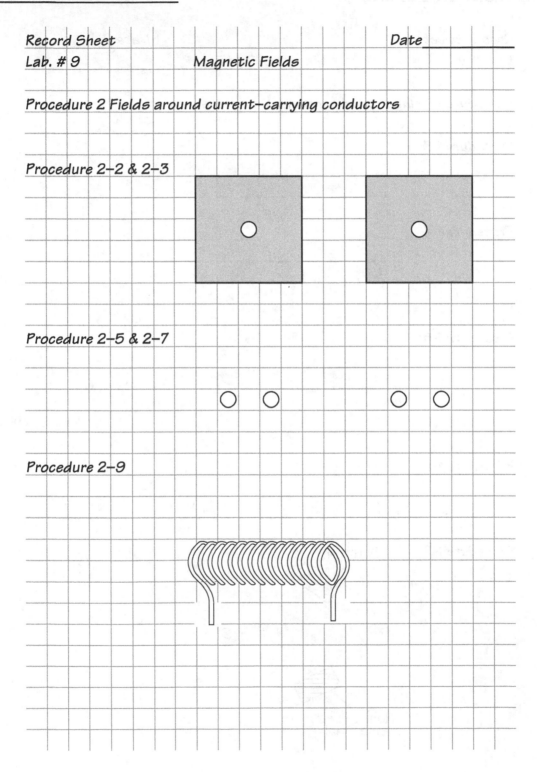

Record Sheet   Date

Lab. # 9   Magnetic Fields

Procedure 2 Fields around current–carrying conductors

Procedure 2–2 & 2–3

Procedure 2–5 & 2–7

Procedure 2–9

LABORATORY INVESTIGATION 10
Electromagnetism

Introduction

A 250 turn solenoid is wound using lacquered copper wire, and a second (30 turn coil) is wound on top of the first coil. A center-zero Galvanometer is used to monitor the voltage induced in the small coil when a dc power supply connected to the solenoid is switched *on* and *off*.

An electromagnet is constructed by winding a coil on to a U-shaped soft iron core. The electromagnet is employed to lift a rectangular iron bar, and the minimum coil current required to hold the bar is investigated.

Equipment

DC Power Supply—(10 V, 1 A)
DC Ammeter—(1 A)
Center-zero Galvanometer—(± 250 μV)
Cylindrical soft iron core—(approximately 10 cm × 1 cm)
Rectangular soft iron bar—(approximately 10 cm × 1 cm × 1 cm)
U-shaped cylindrical soft iron bar—(approximately 10 cm × 0.75 cm)
The U-shaped bar should have its end (or poles) filed flat so that the rectangular bar fits flatly against them.
Lacquered copper wire—(22 gauge or larger, approximately 15 m)

Procedure 1 Solenoid

1-1   Using the lacquered copper wire, closely wind a solenoid of about 250 turns on the cylindrical soft iron core. The coil should be loose enough that the core can be slipped out of it, (see the illustration in Fig. 12-4 in the text book). If necessary, wind two or more layers to comprise 250 turns. Remove the iron core from the solenoid but retain the coil in its cylindrical shape. Adhesive tape can be useful for this.

1-2   Wind a second coil of about 30 turns on top of the solenoid, and tape it to hold its shape and position.

1-3   Clean the insulation off the ends of the copper wire forming the solenoid, and connect the solenoid in series with the dc ammeter to the terminals of the power supply as in Fig. L10-1.

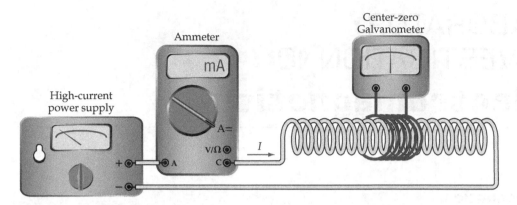

Figure L10-1   Connections of solenoid and secondary coil for investigation of electromagnetic induction.

1-4   Clean the insulation off the ends of the copper wire forming the 30-turn coil, and connect this coil to the terminals of the center-zero Galvanometer.

1-5   Set the power supply output control to zero, then switch the power supply *on* and adjust it to give a current of approximately 1 A through the solenoid.

1-6   Switch the power supply *off* and *on* again, and note the deflection on the Galvanometer. Instead of switching *on* and *off*, try rapidly adjusting the power supply output control to zero and then back again to 1 A. Note the Galvanometer deflection as the control is adjusted.

1-7   Insert the cylindrical soft iron core into the solenoid, then repeat Procedure 1-6.

Procedure 2   Electromagnet

2-1   Using the lacquered copper wire, wind a 200 turn coil on the U-shaped soft iron core.

2-2   Clean the insulation from the ends of the copper wire, then connect the electromagnet coil in series with the dc ammeter to the terminals of the power supply, as in Fig. L10-2.

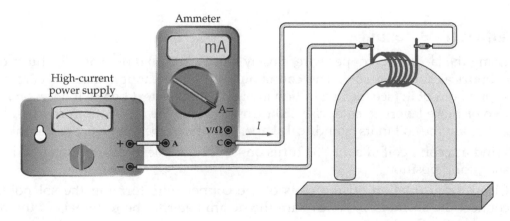

Figure L10-2   Investigation of electromagnet.

2-3 Set the power supply output control to zero, then switch the power supply *on* and adjust it to give a coil current of approximately 1 A.

2-4 Use the electromagnet to lift the rectangular iron bar.

2-5 Remove the iron bar from the magnet, and reduce the coil current to zero.

2-6 Slowly increase the coil current while at the same time trying to lift the bar with the magnet. Note the lowest current level at which the electromagnet becomes powerful enough to lift the iron bar.

2-7 Return the coil current to zero, and tape a thin sheet of paper on surface the rectangular iron bar. Repeat Procedure 2-6.

2-8 Repeat Procedure 2-7 using two sheets of paper.

Analysis

1   Discuss the results of Procedures 1-6 and 1-7. Also calculate the total magnetic flux in the solenoid when the iron core was absent.

2   Discuss the results of Procedures 2-6 through 2-8.

Record Sheet L10

Record Sheet   Date _____

Lab. #10   Electromagnetism

Procedure 1 Solenoid

	Without the iron core	With the iron core
Galvanometer deflection		

Procedure 1 Solenoid

	Direct contact	One sheet of paper	Two sheets of paper
Minimum current to hold the bar			

LABORATORY INVESTIGATION 11
Laboratory-Constructed Analog Instruments

Introduction

A dc ammeter is constructed using a Permanent Magnet Moving Coil (PMMC) instrument with a precision decade resistance box connected as a shunt. The scale reading of the laboratory-constructed ammeter is compared to the measurements made on a commercial ammeter which passes the same current.

A dc voltmeter is constructed using a PMMC instrument with a precision decade resistance box connected as a multiplier resistor. The performance of the laboratory-constructed instrument is compared to a parallel-connected commercial voltmeter.

Equipment

PMMC Instrument with FSD = 100 μA
Decade Resistance Box (R_1)—(1 Ω to 999 kΩ)
Decade Resistance Box (R_S)—(0.1 Ω to 10 kΩ, 100 mA)
DC Ammeter—(10 mA)
DC Voltmeter
DC Power Supply—(0 to 25 V, 10 mA)

Procedure 1 DC Ammeter

1-1 Check the mechanical zero of the 100 μA PMMC instrument (M), and adjust as necessary.

1-2 If analog instruments (ammeter and voltmeter) are used, check the mechanical zeros and adjust as necessary.

1-3 Set the precision decade boxes to maximum resistance then connect the equipment as shown in Fig. L11-1. *Do not connect R_S into the circuit at this time.*

1-4 Switch *on* the power supply and adjust its output to give exactly 1 V on voltmeter V_1.

1-5 Reduce the resistance of R_1 until the PMMC instrument (M) indicates exactly 100 μA. Note the resistance of R_1.

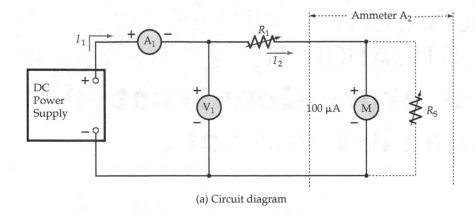

(a) Circuit diagram

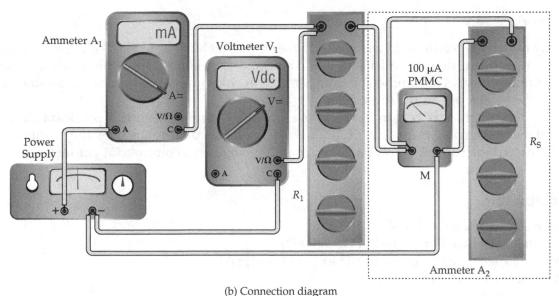

(b) Connection diagram

Figure L11-1 Circuit and connection diagram for testing laboratory-constructed ammeter.

1-6 Calculate the resistance of M from: $R_1 + r_{in} = 1 \text{ V}/100 \text{ } \mu\text{A}$

1-7 Calculate the required value of shunt resistance (R_S) to convert M into a 10 mA ammeter. (See the procedure set out in Example 13-1 in the text book.)

1-8 Set the second decade resistor box to the calculated value of R_S, and connect it as shown dashed in Fig. L11-1(a).

1-9 Adjust the power supply output and R_1 as necessary to give full scale deflection on M [10 mA for the ammeter (A_2) constituted by M and R_S]. Note the exact current indicated on Ammeter A_1.

1-10 Adjust the power supply to reduce the current indicated on A_2 in 2 mA steps. At each step record the exact level of current indicated on ammeter A_1.

Procedure 2 DC Voltmeter

2-1 Adjust the power supply voltage to zero, and rearrange the circuit as shown in Fig. L11-2.

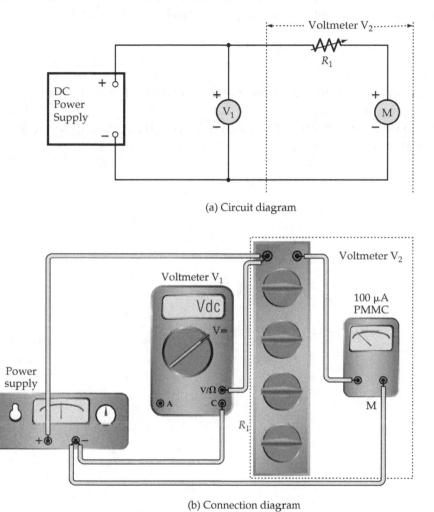

(a) Circuit diagram

(b) Connection diagram

Figure L11-2 Circuit for testing laboratory-constructed voltmeter.

2-2 Using the resistance of r_m already calculated, determine the value of the multiplier resistance required to convert M into a 25 V voltmeter. (See the procedure laid out in Example 13-3 in the text book.)

2-3 Set R_1 to the multiplier resistance value calculated above.

2-4 Increase the power supply output to give full scale deflection on M, (25 V for the voltmeter V_2 constituted by M and R_1). Note the exact voltage indicated on voltmeter V_1.

2-5 Adjust the power supply to reduce the voltage indicated on V_2 in 5 V steps. At each step record the exact level of voltage indicated on voltmeter V_1.

Analysis
1 Sketch the complete circuit of ammeter A_2 constructed in Procedure 2. Calculate the total resistance of A_2.
2 Sketch the complete circuit of voltmeter V_2 constructed in Procedure 2. Calculate the sensitivity of V_2.

Record Sheet L11

Record Sheet   Date _____

Lab. # 11   Laboratory-Constructed Instruments

Procedure 1 DC Ammeter
Procedure 1–5, 1–6, 1–7

	R_1	r_m	R_S

Procedure 1–9, 1–10

	100				
M scale (μA)	10	8	6	4	2
I_2 (mA)					
I_1 (mA)					

Procedure 2 DC Voltmeter
Procedure 2–2

$R_1 =$

Procedure 2–4, 2–5

	100				
M scale (μA)	10	8	6	4	2
V_2 (V)					
V_1 (V)					

LABORATORY
INVESTIGATION 12
Ohmmeter

Introduction

A PMMC instrument, decade resistance box, and power supply are connected together to form a simple ohmmeter circuit. The instrument deflection is investigated when measuring several known resistance values. The circuit is modified to include a zero ohms control, and its resistance scale is again checked. The effects of a decreasing battery voltage and the zero ohms control are investigated.

Equipment

PMMC Instrument with FSD = 100 μA
Three Decade Resistance Boxes—(0 to 100 kΩ, 100 mA)
DC Voltmeter
DC Power Supply—(0 to 25 V, 100 mA)

Procedure

1 Check the mechanical zero of the 100 μA PMMC instrument and adjust as necessary.

2 Set all three decade resistance boxes to maximum resistance, then connect the equipment as illustrated in Fig. L12-1. [Also see Fig. 13-8(a) in the text book.] *Note that R_2 is to be left unconnected at this time.*

3 Set the power supply voltage (E_b) to exactly 3 V, then short terminals A and B together and adjust R_1 to give full scale deflection (FSD) on the PMMC instrument. Record the resistance of R_1.

4 Open circuit A and B to include R_X in the circuit.

5 Adjust R_X to successively give 3/4 FSD, 1/2 FSD, and 1/4 FSD on the PMMC instrument. Record the resistance of R_X at each setting, and compare the resistance values to the ohmmeter scale in Fig. 13-8(b) in the text book.

6 Connect R_2 across the PMMC instrument (as illustrated in Fig. 13-9 in the text book).

7 Short terminals A and B together again, and adjust R_2 to give 1/2 FSD on the PMMC instrument. Record the resistance of R_2.

8 With A and B still shorted, adjust R_1 to return the meter indication to FSD. Record the resistance of R_1.

9 Repeat Procedures 4 and 5.

10 To simulate a reduction in battery terminal voltage, alter the power supply to set E_b to 2.5 V.

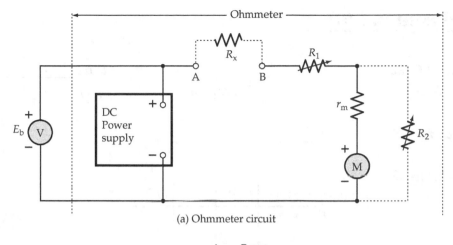

(a) Ohmmeter circuit

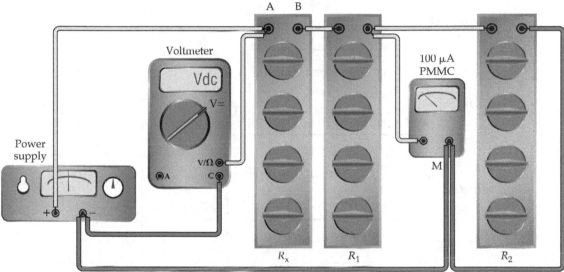

(b) Ohmmeter connection diagram

Figure L12-1 Circuit and connection diagram for laboratory-constructed ohmmeter.

11 Short terminals A and B together once again, and adjust R_2 to give FSD on the PMMC instrument.

12 Repeat Procedures 4 and 5 once more.

Analysis

1 Briefly discuss the performance of the ohmmeter circuit investigated in Procedures 1 through 5. How does this circuit compare to the instrument analyzed in Example 13-5 in the text book?

2 Briefly discuss the performance of the ohmmeter circuit investigated in Procedures 6 through 12. Compare the circuit to the instrument analyzed in Example 13-6 in the text book.

Record Sheet L12

Record Sheet   Date_____

Lab. # 12   Ohmmeter

Procedure 3 $R_1 =$ _____

Procedure 5

Deflection (FSD)	0.75	0.5	0.25
R_X			

Procedure 7, 8

	R_2	R_1

Procedure 9

Deflection (FSD)	0.75	0.5	0.25
R_X			

Procedure 12

Deflection (FSD)	0.75	0.5	0.25
R_X			

LABORATORY INVESTIGATION 13
Wheatstone Bridge

Introduction

A Wheatstone bridge is constructed using three precision decade resistance boxes, a galvanometer, and a power supply. An 'unknown' resistor forms the fourth arm of the bridge. Unknown resistances ranging from 200 Ω to 6.8 kΩ are measured on the bridge.

Equipment

DC Power Supply—(0 to 25 V, 100 mA)
Three Precision Decade Resistance Boxes—(0 to 100 kΩ, 100 mA)
DC Galvanometer—(with sensitivity control)
Three 'unknown' Resistors—(approximately 200 Ω, 1.2 kΩ and 6.8 kΩ)

Procedure

1 Check the mechanical zero of the Galvanometer and adjust as necessary.

2 Connect the equipment as illustrated in Fig. L13-1. (This is the same arrangement as in Fig. 13-20 in the text book.) Use the power supply (set to approximately 5 V) to provide bridge supply voltage E. For R_1, R_3, and R_4, use the decade resistance boxes set to 100 kΩ. For R_2 use one of the unknown resistances.

3 Switch *on* the power supply and adjust R_4 until the Galvanometer indicates zero.

4 Increase the Galvanometer sensitivity and continue (fine) adjustment of R_4 until the Galvanometer indicates zero on its most sensitive range.

5 Record the values of R_1, R_3, and R_4, and calculate the exact value of unknown resistance R_2.

6 Substitute each of the other unknown resistance into the circuit in turn, and repeat Procedures 3, 4, and 5 to determine each resistance value.

7 Alter resistor R_1 to 10 kΩ, then repeat procedures 4, 5, and 6.

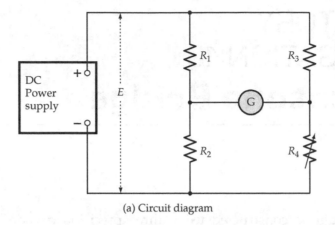

(a) Circuit diagram

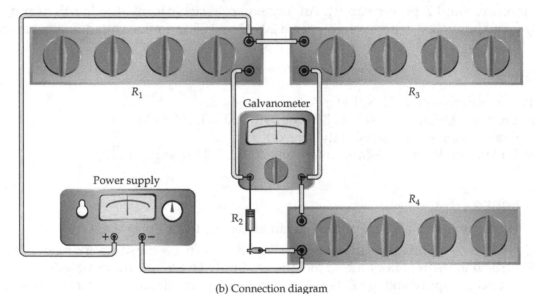

(b) Connection diagram

Figure L13-1 Circuit and connection diagram for a laboratory-constructed Wheatstone bridge.

Analysis

1 Compare the measured resistances to the coded values of each unknown resistances.
2 Look up the specified accuracy of the decade resistance boxes, and calculate the accuracy of each of the three measured resistance values.

Record Sheet L13

Record Sheet

Lab. # 13   Wheatstone Bridge

Date _____

Procedure 5

R_1	R_3	R_4	R_2 coded	R_2 calculated

Procedure 6

R_1	R_3	R_4	R_2 coded	R_2 calculated

Procedure 6

R_1	R_3	R_4	R_2 coded	R_2 calculated

LABORATORY INVESTIGATION 14
DC *RC* Circuit

Introduction

A capacitor and a resistor are connected in series to the terminals of a dc power supply via a switch. Voltmeters are connected to monitor the supply voltage, the resistor voltage, and the capacitor voltage. The times required for the capacitor voltage to reach certain predetermined levels are measured. Similarly, the resistor voltage change is timed. In all cases, the measured times are related to the circuit time constant.

Equipment

DC Power Supply—(0 to 10 V, 100 mA)
DC Voltmeter—(10 V)
Two Electronic Voltmeters (Analog or Digital)—(10 V)
Single Pole Switch
Two 100 μF Capacitors
100 kΩ Resistor—(1/4 watt or larger)
Stop Watch

Procedure

1   Connect the dc power supply, switch, 100 kΩ resistor, and 100 μF capacitor as shown in Fig. L14-1. *Ensure that the capacitor is connected with the correct polarity*.

2   Connect the two electronic voltmeters to monitor the voltage across R and C, and the other voltmeter to monitor the power supply voltage. Set each meter to its 10 V range.

3   With S_1 open, adjust the power supply output to 10 V, and prepare a stop watch to measure the time required to charge the capacitor. (If a stop watch is not available, an ordinary wrist watch with a seconds hand or a digital watch with a seconds read out can be used.)

4   Close S_1 and measure the time for e_c to increase from 0 to 6.3 V, and the time required for it to reach 9 V. Record these times on the record sheet provided.

5   Adjust the power supply voltage to zero, and when all three voltmeters indicate zero open S_1.

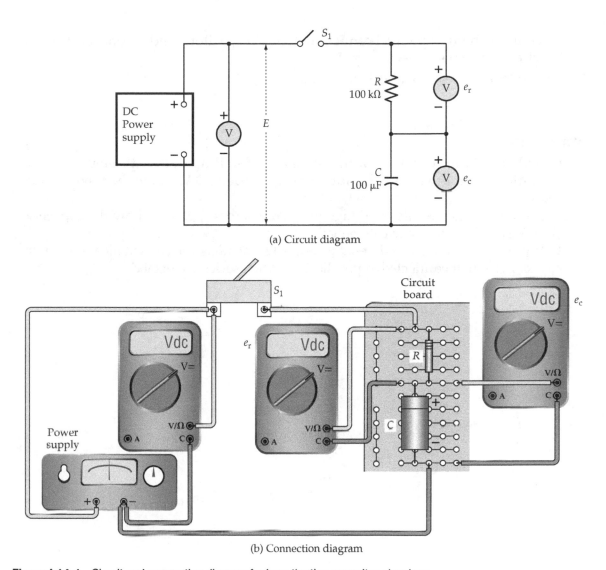

(a) Circuit diagram

(b) Connection diagram

Figure L14-1 Circuit and connection diagram for investigating capacitor charging.

6 Set the power supply voltage to 10 V and close S_1 again. This time observe e_r. Measure and record the time for e_r to arrive at 3.6 V and the time taken for it to arrive at approximately 1 V.

7 With the power supply output still at 10 V open S_1. Record the levels of e_r and e_c.

8 Reverse the polarity of the voltmeter monitoring e_r. Adjust the power supply output to zero then close S_1. Observe the indications of e_r and e_c, and record the time taken for the voltage levels to fall to zero.

9 Replace C with two 100 μF capacitors connected in parallel.

10 Repeat Procedures 3, 4, and 5.

11 Disconnect the two parallel-connected 100 μF capacitors, and reconnect them into the circuit in series.

12 Repeat Procedures 3, 4, and 5 once again.

Analysis

1 Calculate the time constant for the circuit tested in Procedures 1 through 8. Relate the time constant to the time required for e_c to reach 6.3 V, and to the time for e_c to reach 9 V.

2 Using the various levels of voltage measured across e_r explain how the capacitor charge and discharge currents behave.

3 Using the results of Procedures 9 through 12, calculate the capacitance of the two capacitors when connected in parallel and when series-connected.

Record Sheet L14

Record Sheet   Date _____

Lab. # 14   Capacitance

Procedure 4 $R = 100 \text{ k}\Omega, C = 100 \text{ μF}$

	$e_c = 6.3$ V	$e_c = 9$ V
t		

Procedure 6

	$e_c = 3.6$ V	$e_r = 1$ V
t		

Procedure 7

e_r	e_c

Procedure 8

e_r	e_c	time to zero

Procedure 10 $C = 100 \text{ μF} \| 100 \text{ μF}$

	$e_c = 6.3$ V	$e_c = 9$ V
t		

Procedure 12 $C = 100 \text{ μF}$ in series with 100 μF

	$e_c = 6.3$ V	$e_c = 9$ V
t		

LABORATORY INVESTIGATION 15
Oscilloscope

Introduction

Two sinusoidal voltage waveforms are displayed on an oscilloscope, and the effects of the *focus, brightness,* and *trigger* controls are investigated. The peak-to-peak amplitude and time period of each waveform is measured, and the *v/div* and *time/div* controls are investigated. Finally, two out-of-phase waveforms are displayed and their phase difference is measured.

Equipment

Dual-trace Oscilloscope
Two Audio Signal Generators—(Sine Wave)
0.1 μF Capacitor
3.3 kΩ, 1/2 W Resistor

Procedure 1   Setting the Controls

1-1   Switch *on* the oscilloscope and allow it to warm up for a few minutes, then set the instrument controls as follows:

Control	Position
NORMAL – TV	NORMAL
TRIG SOURCE	INT
BA-SWP	SWP
TIME/DIV	1 ms (centre knob calibrated)
TRIGGER LEVEL	AUT

Channel 1 & Channel 2 controls:

AC-DC	AC
VERTICAL DISPLAY	A and B
V/DIV	1
GND Buttons	Released

1-2   Two horizontal lines should now be displayed on the screen. If they are not present, adjust the POSITION and INTENSITY controls as necessary.

1-3 Adjust the FOCUS controls to focus each display to a fine line. Also, alter the INTENSITY controls as necessary to give reasonable bright (but not too bright) displays, and refocus if required.

Procedure 2 Waveform Display

2-1 Connect one of the signal generators to the CHANNEL 1 input of the oscilloscope.

2-2 Switch the signal generator *on* and set its output to a frequency of 250 Hz.

2-3 Adjust the output amplitude of the signal generator to give a display that occupies approximately four vertical divisions on the oscilloscope screen (like the top wave-form in Fig. 17-20 in the text book).

2-4 Connect the other signal generator to the CHANNEL 2 input of the oscilloscope. Set its frequency to 750 Hz and adjust its amplitude to occupy approximately one and a half vertical divisions on the oscilloscope screen. Sketch the two waveforms on the record sheet. *Note that it will be necessary to carefully adjust the frequency of one signal generator to keep the displayed wave from 'sliding off' to one side. This will not be necessary if one of the signal generators is synchronized from the other.*

2-5 Investigate the effect of adjusting the VERTICAL and HORIZONTAL position controls.

Procedure 3 Waveform Measurement

3-1 Without further adjustment, estimate the time period of each input wave in horizontal divisions. Using the TIME/DIV setting, calculate each time period in ms and determine the two input frequencies. *Note that for accurate time measurement the vernier knob of the TIME/DIV control must be in its CAL position.*

3-2 Investigate the effect of adjusting the TIME/DIV control.

3-3 Investigate the effect of adjusting the vernier knob of the TIME/DIV control, then return it to its correct (i.e., calibrated) position.

3-4 Adjust the POSITION, TIME/DIV, and V/DIV controls to expand one of the wave-forms until one cycle approximately fills the screen, as illustrated in Fig. 17-23 in the text.

3-5 Carefully measure the time period and peak-to-peak amplitude of the wave.

Procedure 4 Phase Measurement

4-1 Connect the 0.1 μF capacitor and 3.3 kΩ resistor in series across the terminals of one of a signal generator, as in Fig. L15-1.

4-2 Connect the CHANNEL 1 input of the oscilloscope to monitor the waveform at the signal generator terminals, and the CHANNEL 2 input to monitor the waveform of the voltage across the resistor.

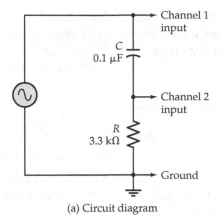

(a) Circuit diagram

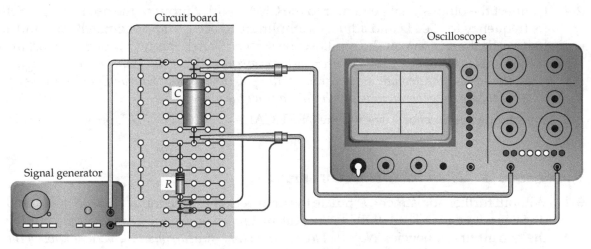

(b) Connection diagram

Figure L15-1 Circuit and connection diagram for phase measurement on an oscilloscope.

4-3 Set the signal generator frequency to 1 kHz, and adjust its amplitude control to give a waveform which approximately fills the top half of the oscilloscope screen.

4-4 Adjust the Channel 2 VERTICAL POSITION and V/DIV controls, to give a display which approximately fills the bottom half of the screen.

4-5 Set the TIME/DIV control to 0.1 ms. Measure the phase difference between the two waveforms as illustrated in Fig. 17-24 and explained in the text.

Analysis

1 Discuss the results of Procedures 1, 2, 3, and 4 in turn.
2 Referring to the controls of the oscilloscope estimate the lowest and highest amplitude waveforms that may be displayed. Also estimate the lowest frequency and highest frequency that may be displayed.

Record Sheet L15

Record Sheet   Date _____

Lab. # 15   Oscilloscope

Procedure 2–4 Waveforms

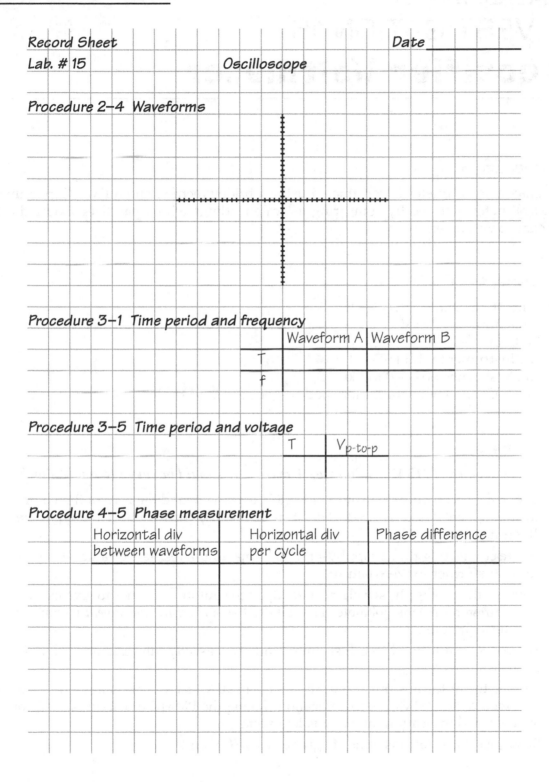

Procedure 3–1 Time period and frequency

	Waveform A	Waveform B
T		
f		

Procedure 3–5 Time period and voltage

T	$V_{p\text{-to-}p}$

Procedure 4–5 Phase measurement

Horizontal div between waveforms	Horizontal div per cycle	Phase difference

71

LABORATORY
INVESTIGATION 16
Rectifier Voltmeter

Introduction

An analog ac voltmeter is constructed using a bridge rectifier and a PMMC instrument. The voltmeter is tested by comparing its scale readings with a parallel-connected commercial ac voltmeter.

Equipment

Isolating Transformer—1:1, 115 V
Autotransformer—115 V
AC Voltmeter—50 V
PMMC Instrument with FSD = 1 mA
Decade Resistance Box—(0 to 100 kΩ), 1 mA
Four Low Current Semiconductor Diodes—(such as IN914)

Procedure

Note that because a 115 V supply is used, extra care should be taken to avoid shock.

1   Check the mechanical zero of the PMMC instrument, and the zero of the voltmeter if an analog voltmeter is used. Adjust as necessary, then connect the equipment as illustrated in Fig. L16-1.

2   Before connecting the ac supply, set R_S to its maximum resistance value, and adjust the autotransformer for zero output.

3   Connect the ac supply, switch *on*, and adjust the autotransformer to give exactly 50 V on voltmeter V_1. This voltmeter is to be used as a *reference instrument* to calibrate the rectifier voltmeter.

4   Reduce R_S until the PMMC instrument indicates exactly full scale deflection. Record the resistance of R_S.

5   Reduce the autotransformer output voltage in 10 V steps as measured on V_1. At each step record the PMMC meter indication. Taking the PMMC full scale as 50 V, convert the recorded indications into equivalent voltages.

6   Re-adjust the autotransformer to give exactly 30 V on V_1.

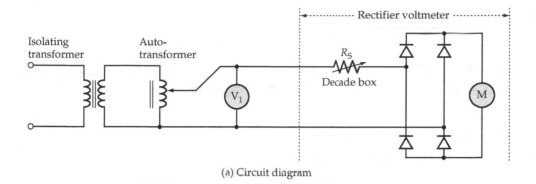

(a) Circuit diagram

(b) Connection diagram

Figure L16-1 Circuit and connection diagram for rectifier voltmeter test.

7   Reduce R_S until the PMMC instrument indicates full scale once again. Record the resistance of R_S.

8   Reduce the autotransformer output in 5 V steps. At each step record the indication on the PMMC instrument. Taking the PMMC full scale as 30 V, convert the recorded indications into equivalent voltages.

Analysis

1   Analyze the 50 V rectifier voltmeter investigated in Procedures 3 through 6 to determine the meter resistance R_m. (Refer to Example 17-8 in the text book.)

2   Using the calculated value of R_m, determine the required multiplier resistance R_S to convert the rectifier voltmeter to 30 V FSD. Compare this to the R_S value determined in Procedure 8.

Record Sheet L16

Record Sheet   Date _____

Lab. # 16   Rectifier Voltmeter

Procedure 4

$R_s =$ _____

Procedure 5

V_1 reading (V)	50	40	30	20	10
PMMC current (mA)					
Equivalent measured voltage (V)					

Procedure 7

$R_s =$ _____

Procedure 8

V_1 reading (V)	30	25	20	10	5
PMMC current (mA)					
Equivalent measured voltage (V)					

LABORATORY INVESTIGATION 17
AC *RL* Circuit

Introduction

A sinusoidal signal is applied to a series resistive-inductive circuit, and the voltages developed across the resistor and inductor are investigated for amplitude and phase relationship to the input. A square wave input is next applied, and the inductor and resistor voltages are again investigated.

Equipment

Dual-trace Oscilloscope
Low Frequency Signal Generator—(Sine and Square Waves)
4 H Inductor—(winding resistance less than 500 Ω)
4.7 kΩ Resistor—(1/4 W or larger)

Procedure

1 Connect the 4 H inductor and 4.7 kΩ resistor to the signal generator as illustrated in Fig. L17-1. [This is the same as in Fig. 19-15(a) in the text book.]
2 Connect the oscilloscope to monitor the input voltage (E) and the inductor voltage (V_L). The two ground terminals of the oscilloscope inputs should be connected to lower terminal of the inductor in the circuit diagram; that is, to the L terminal connected to the signal generator.
3 Switch *on* the signal generator and set it to give a sine wave output with a frequency of 250 Hz. Adjust the signal amplitude to give waveforms which approximately fill half the oscilloscope screen.
4 Set the oscilloscope to trigger positively on the input waveform, and adjust the time base to display approximately one cycle of each waveform.
5 Measure the waveform amplitudes and phase difference (as explained in Section 17-6 in the text book). Enter the measured quantities on the record sheet.
6 Reconnect the oscilloscope to monitor E and V_R. This time connect the two ground terminals of the oscilloscope to the top of R in the circuit diagram.
7 Repeat Procedure 5.
8 Switch the signal generator output to square wave.
9 Observe the waveforms of E and V_R on the oscilloscope. Carefully measure the waveform amplitudes and note their phase relations. Record the measured quantities and sketch the waveforms on the record sheet.

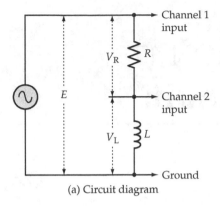

(a) Circuit diagram

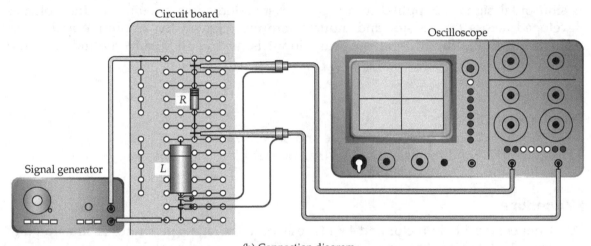

(b) Connection diagram

Figure L17-1 Circuit and connection diagram for *RL* circuit investigation.

10 Reconnect the oscilloscope to monitor E and V_L once more. Again connect the grounded input terminals to the lower terminal of L in the circuit diagram.

11 Sketch the waveforms, and carefully measure and record the amplitude and phase relationships of E and V_L.

Analysis

1 Use the waveform amplitudes and phase relationships determined during Procedures 1 through 7 to sketch the waveforms of I, V_R, V_L, and E in the form illustrated in Fig. 19-15(b) in the text book.

2 Sketch a phasor diagram for the *RL* circuit, using the measured values of V_L, E, V_R, and ø.

3 Briefly explain the waveforms obtained in Procedures 9 and 11.

Record Sheet L17

Record Sheet   **Date** _____

Lab. # 17   AC RL Circuit

Procedure 5   Sine wave input

	$E_{(p\text{-to-}p)}$	$V_{L(p\text{-to-}p)}$	Phase difference

Procedure 7

	$E_{(p\text{-to-}p)}$	$V_{R(p\text{-to-}p)}$	Phase difference

Procedure 9   Square wave input

$E_{(p\text{-to-}p)}$	$V_{R(p\text{-to-}p)}$

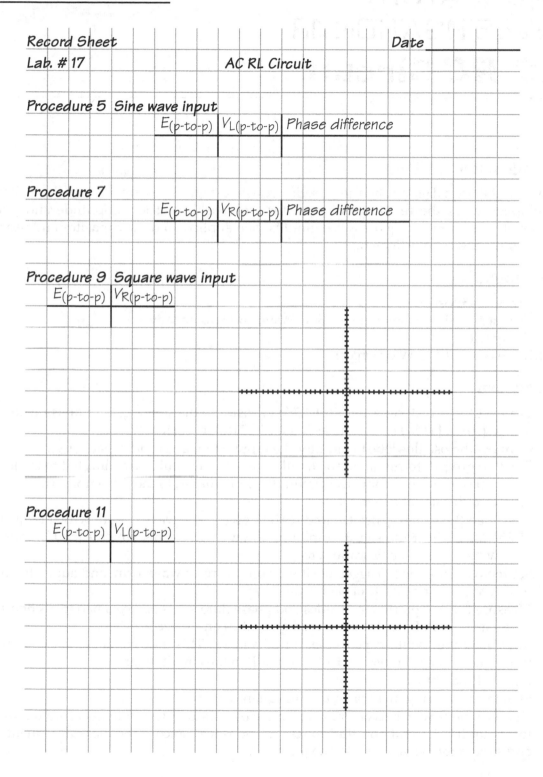

Procedure 11

$E_{(p\text{-to-}p)}$	$V_{L(p\text{-to-}p)}$

LABORATORY INVESTIGATION 18
AC *RC* Circuit

Introduction

A sinusoidal signal is applied to a series resistive-capacitive circuit, and the voltages developed across the resistor and capacitor are investigated for amplitude and phase relationship to input. A square wave input is then applied, and the capacitor and resistor voltages are again investigated.

Equipment

Dual-trace Oscilloscope
Low Frequency Signal Generator—(Sine Wave and Square Wave)
0.1 μF Capacitor
4.7 kΩ Resistor—(0.25 W or larger)

Procedure

1   Connect the 0.1 μF capacitor and 4.7 kΩ resistor to the signal generator as illustrated in Fig. L18-1. [This is the same as in Fig. 19-21(a) in the text book.]
2   Connect the oscilloscope to monitor the input voltage (*E*) and capacitor voltage (*V*$_C$). The two ground terminals of the oscilloscope input should be connected to the lower terminal of the capacitor in the circuit diagram; that is, to the capacitor terminal connected to the signal generator.
3   Switch *on* the signal generator and set it to give a sine wave output with a frequency of 250 Hz. Adjust the signal amplitude to give waveforms which each fill approximately half of the oscilloscope screen.
4   Set the oscilloscope to trigger positively on the input waveform, and adjust the time base to display one cycle of each waveform.
5   Measure the waveform amplitudes and phase differences as explained in Section 17-6 in the text book. Enter the measured quantities on the record sheet.
6   Reconnect the oscilloscope to measure *E* and *V*$_R$. This time connect the two ground terminals of the oscilloscope to the top of *R* in the circuit diagram.
7   Repeat Procedure 5.
8   Switch the signal generator output to square wave.
9   Observe the waveforms of *E* and *V*$_R$ on the oscilloscope. Carefully measure the waveform amplitudes and note their phase relationships. Record the measured quantities and sketch the waveforms on the record sheet.

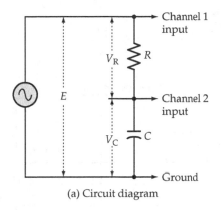

(a) Circuit diagram

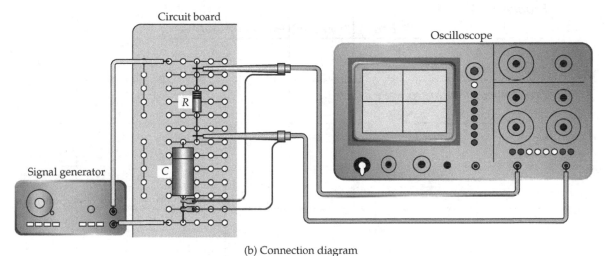

(b) Connection diagram

Figure L18-1   Circuit and connection diagram for *RC* circuit investigation.

10   Reconnect the oscilloscope to monitor E and V_C once again. Again connect the grounded inputs to the lower terminal of C in the circuit diagram.

11   Carefully measure the amplitudes and phase relationships of E and V_C, and sketch the waveforms.

Analysis

1   Use the waveform amplitudes and phase relationships determined during procedures 1 through 7 to sketch the waveforms of I, V_R, V_C, and E in the form illustrated in Fig. 19-21(b) in the text book.

2   Sketch a phasor diagram for the *RC* circuit using the measured values of V_C, E, V_R, and ø.

3   Briefly explain the waveforms obtained in Procedures 9 and 11.

79

Record Sheet L18

Record Sheet   Date _____

Lab. # 18   AC RC Circuit

Procedure 5 | Sine wave input

	$E_{(p\text{-}to\text{-}p)}$	$V_{C(p\text{-}to\text{-}p)}$	Phase difference

Procedure 7

	$E_{(p\text{-}to\text{-}p)}$	$V_{R(p\text{-}to\text{-}p)}$	Phase difference

Procedure 9 | Square wave input

$E_{(p\text{-}to\text{-}p)}$	$V_{R(p\text{-}to\text{-}p)}$

Procedure 11

$E_{(p\text{-}to\text{-}p)}$	$V_{C(p\text{-}to\text{-}p)}$

80

LABORATORY INVESTIGATION 19
Series and Parallel Impedance Circuits

Introduction

A series impedance circuit is constructed and supplied from a signal generator. An oscilloscope is used to measure the voltage at various points in the circuit, and to determine the phase angle of each voltage with respect to the supply. A parallel impedance circuit is also constructed and supplied from a signal generator. The oscilloscope is used to monitor the voltage drop across the resistive component of each impedance, and to measure the phase angle of each voltage with respect to the supply.

Equipment

Dual-trace Oscilloscope
Low Frequency Signal Generator—(Sine Wave)
Resistors: $R_1 = 1\ k\Omega$, $R_2 = 820\ \Omega$ (0.25 W or larger)
Inductor: $L_1 = 0.1\ H$
Capacitor: $C_2 = 0.2\ \mu F$

Procedure 1   Voltage Divider

1-1   Using an ohmmeter measure the winding resistance of inductor L_1. Enter the measured value on the record sheet.

1-2   Connect the signal generator and components as illustrated in Fig. L19-1.

1-3   Connect the oscilloscope to monitor the input voltage (V_i) and the voltage across resistor R_2.

1-4   Adjust the signal generator to give $V_i = 10\ V$ peak-to-peak and $f = 500\ Hz$.

1-5   Carefully measure and record the peak-to-peak value of V_{R2} and its phase angle ($\o_2$) with respect to V_i.

1-6   Reconnect the CHANNEL 2 input of the oscilloscope to terminal C. Measure and record V_2 and its phase angle ($\o_1$) with respect to V_i.

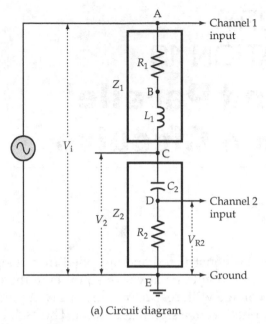

(a) Circuit diagram

(b) Connection diagram

Figure L19-1 Circuit and connection diagram for series impedance circuit investigation.

1-7 Reconnect the CHANNEL 1, CHANNEL 2, and the ground connections of the oscilloscope to terminals A, C, and E as illustrated in Fig. L19-2. Measure and record the peak-to-peak value of V_3 and its angle ($\varnothing_3$) with respect to V_i.

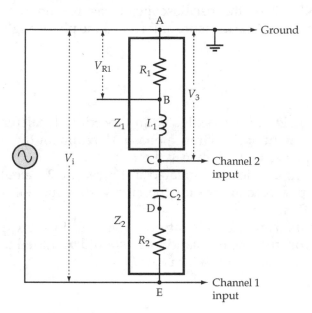

Figure L19-2   Oscilloscope connections for measuring V_{R1} and V_3.

1-8 Reconnect the oscilloscope CHANNEL 1 input to terminal B. Measure and record the peak-to-peak value of V_{R1} and its phase angle ($\varnothing_4$) with respect to V_i.

Procedure 2   AC Current Divider

2-1 Reconnect R_1, L_1, R_2, C_2, and the signal generator as in Fig. L19-3 (and as illustrated in Fig. 20-10 in the text book).

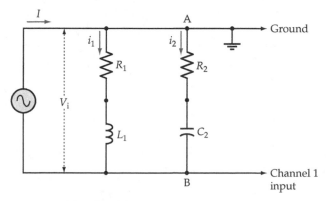

Figure L19-3   Circuit for ac current divider investigation.

83

2-2 Connect CHANNEL 1 of the oscilloscope to monitor supply voltage V_i, using terminal A as ground, (that is, ground the common point of R_1 and R_2).

2-3 Adjust the signal generator to give $V_i = 10$ V peak-to-peak and $f = 500$ Hz.

2-4 Using CHANNEL 2 of the oscilloscope, measure and record the peak-to-peak values of V_{R1} and V_{R2}. Also measure and record the phase angles of V_{R1} and V_{R2} with respect to V_i.

Analysis

1 For the AC voltage divider investigated in Procedure 1, calculate the current level from the measurement of V_{R2}. Draw a phasor diagram for the circuit showing V_i, I, V_2, V_{R2} and the phase angles between them.

2 For the AC current divider investigated in Procedure 2, calculate the values of i_1 and i_2. Draw a phasor diagram for the circuit showing V_i, i_1, i_2, and the phase angles between them.

3 Mathematically analyze each circuit and compare to the experimentally determined results. Note that the resistance of L_1 should be added to R_1 to give the total value of resistance in series with inductor L_1.

Record Sheet L19

Record Sheet   Date _____

Lab. # 19   Series and Parallel Impedance Circuits

Procedure 1 AC voltage divider

$R_L =$ _____

Procedure 1–5, 1–6

	$V_{R2(p\text{-}to\text{-}p)}$	$\emptyset_2$	$V_{2(p\text{-}to\text{-}p)}$	$\emptyset_1$

Procedure 1–7, 1–8

	$V_{3(p\text{-}to\text{-}p)}$	$\emptyset_3$	$V_{R1(p\text{-}to\text{-}p)}$	$\emptyset_4$

Procedure 2 AC current divider

Procedure 2–4

	$V_{R1(p\text{-}to\text{-}p)}$	$\emptyset_1$	$V_{R2(p\text{-}to\text{-}p)}$	$\emptyset_2$

LABORATORY
INVESTIGATION 20
Power in AC Circuits

Introduction

A series resistive-inductive circuit is constructed and connected to a signal generator. The circuit current and its phase angle are monitored by means of an oscilloscope connected across an additional series-connected resistor. Capacitors are connected in parallel with the RL circuit to alter the current and phase angle. The process is repeated with a parallel RL circuit. The results are analyzed to determine apparent power, true power, reactive power, and power factor for each circuit condition.

Equipment

Dual-trace Oscilloscope
Sinusoidal Signal Generator—(600 Hz, 10 V output)
AC Voltmeter—10 V
AC Ammeter—20 mA
Inductor L_1—0.3 H, R <50 Ω
Capacitor C_1—0.1 μF, 20 V
Capacitor C_2—0.05 μF, 20 V
Resistor R_1—5 Ω, 1/4 W
Resistor R_2—1 kΩ, 1/4 W

Procedure

1   Construct the circuit as illustrated in Fig. L20-1. *Do not connect the capacitor at this time.*

2   Switch *on* the signal generator and adjust it to give a circuit input voltage of $V = 10$ V at a frequency $f = 600$ Hz.

3   Use the oscilloscope to measure the phase difference between V_{R1} and V, (between I and V). Record the indicated levels of supply voltage (V) and current (I), and the measured phase difference (ø).

4   Connect capacitor C_1 (0.1 μF) in parallel with R_2 and L_1 as illustrated in Fig. L20-1, then repeat Procedure 3.

5   Replace C_1 with C_2 (0.05 μF), then repeat Procedure 3 once again.

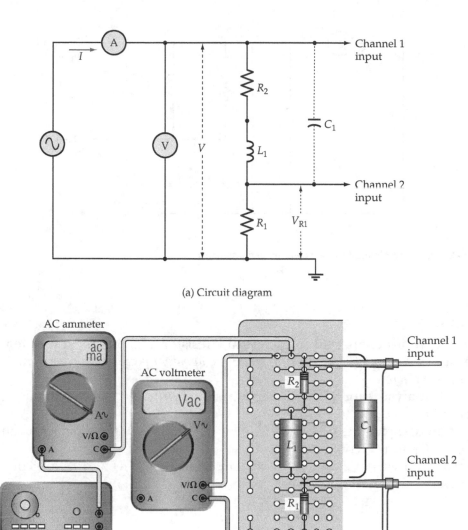

(a) Circuit diagram

(b) Connection diagram

Figure L20-1 Series *RL* circuit for ac power investigation.

6 Reconnect the circuit with R_2 and L_1 in parallel as shown in Fig. L20-2. *Do not connect the capacitor at this time.*

7 Repeat the measurements detailed in Procedure 3.

8 Connect capacitor C_1 in parallel with R_2 and L_1 as illustrated in Fig. 2, then repeat Procedure 3.

9 Remove C_1 and replace it with C_2, then repeat Procedure 3 once more.

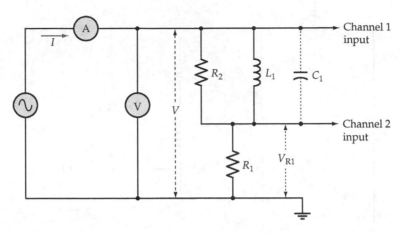

Figure L20-2   Parallel *RL* circuit for ac power investigation.

Analysis

1 For the series-connected inductor and resistor without the capacitor, use the measured values of V, I, and ø to calculate *apparent power, true power, reactive power,* and *power factor.*

2 Repeat Analysis 1 for the measurements made with C_1 connected in the circuit, and then with C_2 connected.

3 Accurately sketch the power triangle for the circuit without a capacitor, and for the cases of C_1 and C_2 connected.

4 Repeat Analysis 1, 2, and 3 for the circuit in which R_2 and L_1 are connected in parallel.

Record Sheet L20

Record Sheet   Date _____

Lab. # 20   Power in AC Circuits

	V	I	Ø
Procedure 3			
" 4			
" 5			
" 7			
" 8			
" 9			

LABORATORY
INVESTIGATION 21
Series Resonance

Introduction

A series *RLC* circuit is constructed and supplied from a signal generator. An oscilloscope is used to monitor the supply voltage and the voltage across R, C, and L in turn. The signal frequency is adjusted for resonance, and the component voltages are noted. The process is repeated at each of several signal frequencies to obtain a table of values from which graphs of V_R, V_C, and V_L may be plotted versus frequency.

Equipment

Dual-trace Oscilloscope
Sinusoidal Signal Generator—200 kHz, 5 V output
Inductor—10 mH, winding resistance less than 30 Ω
Decade Capacitance—(100 pF to 0.01 μF),
    or precision 0.001 μF capacitor
Ohmmeter
Resistor—330 Ω, 1/4 W or larger.

Procedure

1   Use the ohmmeter to measure the inductor winding resistance. Note this value on the record sheet.

2   Connect the resistor, inductor, capacitor, and signal generator as shown in Fig. L21-1.

3   Set the capacitor to 0.001 μF, and the signal frequency to 50 kHz.

4   Connect the oscilloscope to monitor the signal generator voltage (V), and the resistor voltage (V_R). For low oscilloscope input capacitance, a 10:1 probe should be used for the oscilloscope connection to R. *Note that the two grounded input terminals of the oscilloscope must be connected to the bottom terminal in the circuit diagram; that is, to the common junction of the signal generator and R.*

5   Adjust the signal generator voltage to exactly 5 V peak-to-peak, and then adjust the signal frequency (by small amounts) to give the largest obtainable voltage across R. Use the oscilloscope to determine the exact frequency (f_r). Enter the measured frequency on the record sheet.

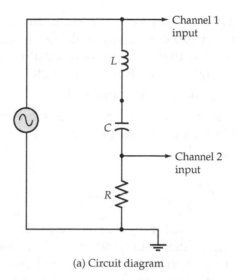

(a) Circuit diagram

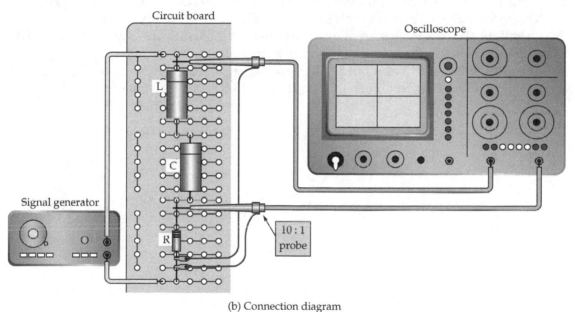

(b) Connection diagram

Figure L21-1   Circuit and connection diagram for series resonance investigation.

6   Check that V is exactly 5 V peak-to-peak, then measure and record the peak-to-peak value of V_R.

7   Without altering the supply voltage and frequency, interchange C and R. The oscilloscope should remain connected so that it now monitors the input voltage and the capacitor voltage V_C, and so that its grounded input terminals are connected to the common junction of the signal generator and capacitor.

8   Measure and record the peak-to-peak value of V_C.

9   Without altering the supply voltage or frequency, interchange L and C. Once again, the oscilloscope should remain connected so that it now monitors the input voltage and inductor voltage, and so that its grounded input terminals are connected to the common junction of the signal generator and inductor.

10 Measure and record the peak-to-peak value of V_L.

11 Set the signal generator to the following frequencies in turn: $0.25 f_r$, $0.5 f_r$, $0.8 f_r$, $1.25 f_r$, $2 f_r$, $4 f_r$. At each input frequency repeat Procedures 6 through 9. Record your results in tabular form on the record sheet.

Analysis

1   Plot graphs of V_R, V_C, and V_L versus frequency (as in Fig. 23-7 in the text book). Note that the frequency values $0.25 f_r$, $0.5 f_r$, f_r, $2 f_r$, and $4 f_r$ should be equally spaced on the horizontal axis to give a logarithmic base.

2   From V_R and R, calculate the current level for each frequency. Plot I to a logarithmic frequency base (to give a graph like that in Fig. 23-5 in the text book).

3   Calculate the resonance frequency for the circuit using the values of L and C. Compare this to the measured resonance frequency.

4   From the measured values of V_L, V_C, and V_R at resonance, estimate the circuit Q. Also, calculate the circuit Q using Equations 23-8 and 23-9 in the text book.

Record Sheet L21

Record Sheet   Date _____

Lab. # 21   Series Resonance

Procedure 1 Winding resistance

$R_L =$ _____

Procedure 5 to 10

f_r	$V_{(p\text{-to-}p)}$	$V_{R(p\text{-to-}p)}$	$V_{C(p\text{-to-}p)}$	$V_{L(p\text{-to-}p)}$

Procedure 11

f_r	$V_{R(p\text{-to-}p)}$	$V_{C(p\text{-to-}p)}$	$V_{L(p\text{-to-}p)}$

LABORATORY INVESTIGATION 22
Parallel Resonance

Introduction

A parallel *RLC* circuit is constructed and supplied from a signal generator. The signal current and the inductor current are monitored on an oscilloscope, and the frequency is adjusted to obtain resonance. The effect of a damping resistance connected across the parallel resonant circuit is investigated. By setting the signal to several frequencies above and below resonance, a table is obtained for plotting a graph of current versus frequency.

Equipment

Dual-trace Oscilloscope
Sinusoidal Signal Generator—10 kHz to 400 kHz, 10 V output.
Inductor—10 mH, resistance <100 Ω
Capacitor—0.001 μF, 20 V
Ohmmeter
Resistors—100 Ω, 1 kΩ, 33 kΩ, all 0.25 W.

Procedure

1 Use the ohmmeter to measure the inductor winding resistance. Note the resistance value on the record sheet.

2 Connect the equipment as illustrated in Fig. L22-1. The oscilloscope grounded input terminals should be connected to the common ground point as shown, and 10:1 probes should be used to give low oscilloscope input capacitance. *Do not include the 33 kΩ resistor at this time.*

3 Switch *on* the signal generator and adjust its output to give $V = 8$ V peak-to-peak and $f \approx 50$ kHz on CHANNEL 1.

4 Carefully adjust the frequency to obtain *minimum* voltage across R_1. (Maximum impedance of the parallel *LC* circuit resulting in minimum supply current.)

5 Measure the resonance frequency f_r from the displayed waveform, and record the measured quantity. The signal generator amplitude control should be adjusted as necessary to maintain $V = 8$ V peak-to-peak.

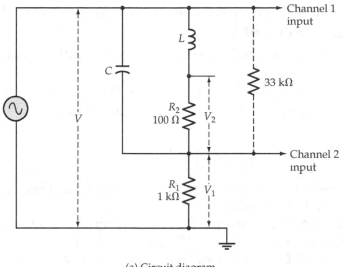

(a) Circuit diagram

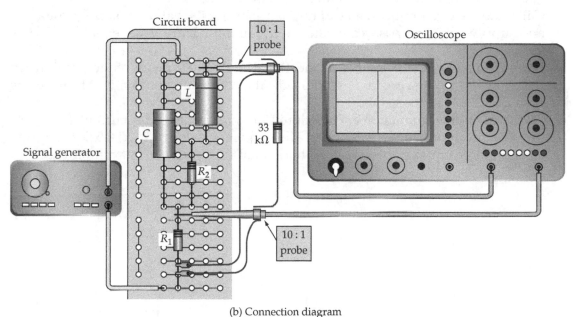

(b) Connection diagram

Figure L22-1 Circuit and connection diagram for parallel resonance investigation.

6 Carefully measure and record the peak-to-peak level of voltage V_1.

7 Temporarily short-circuit from R_1, and reconnect the CHANNEL 2 input of the oscilloscope to monitor voltage V_2.

8 Carefully measure and record the peak-to-peak level of voltage V_2.

9 Remove the short-circuit from R_1 and reconnect the CHANNEL 2 input to monitor voltage V_1.

10 Connect the 33 kΩ resistor in parallel with L and R_2, as shown in Fig. L22-1(a).
11 Repeat Procedures 6 through 9.
12 Remove the 33 kΩ resistor and reconnect CHANNEL 2 of the oscilloscope to monitor V_1 once again.
13 Set the signal frequency in turn to: $0.25\,f_r$, $0.5\,f_r$, $0.8\,f_r$, f_r, $1.25\,f_r$, $2\,f_r$, $4\,f_r$. Measure and record the peak-to-peak level of V_1 at each signal frequency.

Analysis

1 Using the values of L and C, calculate the circuit resonance frequency and compare it to f_r as measured in Procedure 5.
2 From V_1 and V_2 determined during Procedures 6 and 8, calculate the circuit current I and the inductor current I_L; $I = V_1/R_1$, and $I_L = V_2/R_2$.
3 Calculate the circuit Q from the levels of I and I_L at resonance. Also, calculate the circuit Q using Equations 23-8 and 23-9 in the text book.
4 Calculate the circuit impedance at resonance, using the values of supply voltage V and supply current I. Also, calculate the circuit impedance using equation 23-20 in the text book.
5 From the results of Procedures 10 and 11, calculate the Q factor of the circuit when the 33 kΩ damping resistor is employed. Also, calculate the circuit Q factor using Equation 23-25 in the text book.
6 From each V_1 level obtained in Procedure 13, calculate the supply current; $I = V_1/R_1$. Plot the graph of I versus frequency. The frequency values $0.25\,f_r$, $0.5\,f_r$, f_r, $2\,f_r$, and $4\,f_r$ should be equally spaced on the horizontal axis to give a logarithmic base.

Record Sheet L22

Record Sheet

Lab. # 22   Parallel Resonance

Date _____

Procedure 1   Winding resistance

$R_L =$ _____

Procedure 5 to 8

f_r	V_1	V_2

Procedure 11   33 kΩ damping resistor

f_r	V_1	V_2

Procedure 13   Frequency response

f_r	V_1	$I = V_1/R_1$

LABORATORY
INVESTIGATION 23
Low-Pass and High-Pass Filters

Introduction

A sinusoidal signal is applied to a low-pass *RC* filter circuit. The input and output are monitored on an oscilloscope, and the signal frequency is adjusted in steps to obtain a table of input and output voltage levels for plotting the filter gain/frequency response graph. The filter phase shift is also measured at each signal frequency for plotting the phase/frequency response. A high-pass *RC* filter is similarly investigated.

Equipment

Audio Signal Generator—(Sine Wave)
Oscilloscope
Resistors: 1.5 kΩ, 820 Ω, 12 kΩ, (0.25 W or larger)
Capacitors: 3300 pF, 0.1 μF

Procedure 1   Low-Pass Filter

1-1   Using $R = (1.5 \text{ k}\Omega + 820 \text{ }\Omega$ in series), and $C = 0.1$ μF, construct the low-pass *RC* filter circuit as shown in Fig. L23-1.

1-2   Connect an oscilloscope to monitor the input and output waveforms of the filter circuit, as illustrated.

1-3   Adjust the input signal to 1 V peak-to-peak with an approximate frequency of 1 kHz.

1-4   Carefully adjust the signal frequency to give $V_o = 0.707 \ V_i$. Enter the signal frequency as f_c on the laboratory record sheet.

1-5   Check that the input amplitude is 1 V peak-to-peak, then measure and record the output voltage amplitude and the phase difference between the input and output waveforms.

1-6   Successively adjust the input to the following frequencies, and repeat procedure 1-5 at each frequency: $0.125 \ f_c$, $0.25 \ f_c$, $0.5 \ f_c$, f_c, $2 \ f_c$, $4 \ f_c$, $8 \ f_c$.

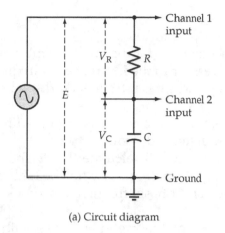

(a) Circuit diagram

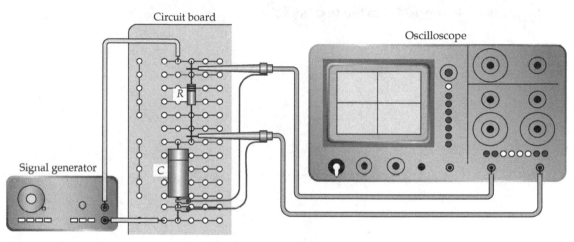

(b) Connection diagram

Figure L23-1   Circuit and connection diagram for *RC* filter investigation.

Procedure 2   High-Pass Filter

2-1   Substitute $R = 12\,\text{k}\Omega$ and $C = 3300\,\text{pF}$ in the filter circuit, and interchange the R and C positions to convert the circuit into a high-pass filter.

2-2   Adjust the input signal to 1 V peak-to-peak with an approximate frequency of 16 kHz.

2-3   Carefully adjust the signal frequency to give $V_o = 0.707\,V_i$. Enter the signal frequency as f_c on the laboratory record sheet.

2-4   Check that the input amplitude is 1 V peak-to-peak, then measure and record the output voltage amplitude and the phase difference between the input and output waveforms.

2-5   Successively adjust the input to the following frequencies, and at each frequency repeat Procedure 2-4: $0.125\,f_c$, $0.25\,f_c$, $0.5\,f_c$, f_c, $2\,f_c$, $4\,f_c$, $8\,f_c$.

Analysis

1 From the results of Procedure 1-6 for the low-pass filter calculate the decibel change in output voltage with respect to the input at each frequency.

2 Plot the gain/frequency and phase/frequency response graphs for the low-pass filter.

3 Compare the performance of the low-pass filter with the calculated results of Examples 24-2 and 24-4 in the text book.

4 From the results of Procedure 2-5 calculate the decibel change in output voltage with respect to the input at each frequency for the high-pass filter.

5 Plot the gain/frequency and phase/frequency response graphs for the high-pass filter.

6 Compare the performance of the high-pass filter with the calculated results of Examples 24-5 and 24-6 in the text book.

Record Sheet L23

Record Sheet   Date _____

Lab. # 23   Low-Pass and High-Pass Filters

Procedure 1–4, 1–5, 1–6

	$f_{(HZ)}$	$Vo_{(p\text{-}to\text{-}p)}(V)$	$\emptyset$	Gain (dB)
0.125 f_c				
0.25 f_c				
0.5 f_c				
f_c				
2 f_c				
4 f_c				
8 f_c				

Procedure 2–3, 2–4, 2–5

	$f_{(HZ)}$	$Vo_{(p\text{-}to\text{-}p)}(V)$	$\emptyset$	Gain (dB)
0.125 f_c				
0.25 f_c				
0.5 f_c				
f_c				
2 f_c				
4 f_c				
8 f_c				

LABORATORY
INVESTIGATION 24
Transformers

Introduction

A transformer with two secondary windings is supplied with 115 V via an isolating transformer and an autotransformer. The secondary voltages are investigated with the secondary windings open-circuited, connected series-opposing, and connected series-aiding. The primary voltage, primary current, and secondary voltage are measured with the secondary loaded. The transformer turns ratio, voltage regulation, and efficiency are determined.

Equipment

Dual-trace Oscilloscope
Isolating Transformer—1:1, 115 V
Autotransformer—115 V
Resistors—$R_1 = 100\ \Omega$, 0.25 W, R2 = 220 Ω, 5 W
Transformer T_1—primary = 115 V,
   two secondaries $\approx$ 15 V each, secondary load $\approx$ 100 mA
AC Voltmeter—150 V (V_1)
AC Voltmeter—50 V (V_2)

Procedure

Note that because a 115 V supply is used, extra care should be taken to avoid shock.

1   Connect the equipment as illustrated in Fig. L24-1.

2   Switch *on* the 115 V supply and adjust the autotransformer to give a transformer primary voltage of $V_P = 115$ V (as indicated on voltmeter V_1). Carefully measure and record the level of the secondary voltage V_{S1}.

3   Reconnect voltmeter V_2 to monitor V_{S2}. Record the level of V_{S2}.

4   Switch *off* the supply, then connect the secondary windings of T_1 series-opposing, as illustrated in Fig. L24-2(a).

5   Connect voltmeter V_2 to monitor the total output voltage V_S of transformer T_1.

6   Switch *on* the supply and check that V_P is still 115 V. Adjust the autotransformer if necessary. Record the total transformer output voltage.

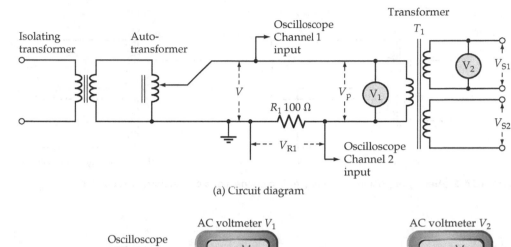

(a) Circuit diagram

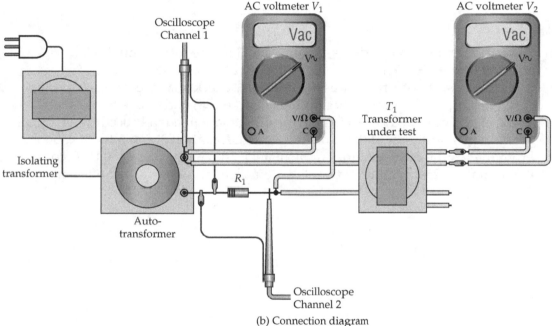

(b) Connection diagram

Figure L24-1 Circuit and connection diagram for transformer investigation.

7   Switch *off* the supply, then reconnect the secondary windings of T_1 in series-aiding, as illustrated in Fig. L24-2(b).

8   Repeat Procedure 6.

9   Switch *off* the supply, then connect the 220 Ω resistor (R_2) across the output terminals of T_1, (with the outputs still connected series-aiding).

10 Switch *on* the supply once again.

11 Measure and record the peak-to-peak level of V_{R1} as displayed on the oscilloscope. Also, measure the phase angle of V_{R1} with respect to input voltage V, and note the transformer output voltage V_S.

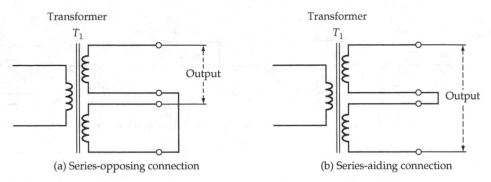

(a) Series-opposing connection   (b) Series-aiding connection

Figure L24-2   Series-opposing and series-aiding connections of transformer secondary.

Analysis

1   From the results of Procedures 2 and 3, calculate the turns ratio; $N_P/N_{S1} = V_P/V_{S1}$ and $N_P/N_{S2} = V_P/V_{S2}$.

2   From the results of Procedures 4 through 8, check that the series-opposing and series-aiding connections give outputs of $(V_{S1} - V_{S2})$ and $(V_{S1} + V_{S2})$ respectively.

3   Using $V_{o(NL)}$ obtained in Procedure 8 and $V_{o(FL)}$ obtained in Procedure 11, calculate the voltage regulation of the transformer. Use Equation 25-14 in the text book.

4   From the results of Procedure 11, calculate: $P_o = V_{S2}^2/R_2$, and $P_i = V_P I_P \cos \o_P$.

5   Use Equation 25-15 in the text book to determine the transformer efficiency.

Record Sheet L24

Record Sheet
Lab. # 24   Transformers   Date _____

Procedure 2 & 3 No-load voltages

V_p	V_{S1}	V_{S2}

Procedure 6 & 8 Series-connected outputs

$V_{S(series-opposing)}$	$V_{S(series-aiding)}$

Procedure 11 Loaded performance

$V_{R1(p-to-p)}$	$V_{(rms)}$	$\emptyset$	$V_{S(rms)}$

LABORATORY INVESTIGATION 25
Three-Phase Y-Connected Loads

Introduction

Wye-connected loads with a three-phase supply are investigated. The loads are (a) balanced resistive, (b) unbalanced resistive, (c) unbalanced impedance. The circuits are analyzed, and the calculated and measured values are compared.

Equipment

Three Decade Resistance Boxes—500 Ω, 500 mA
Capacitor—10 μF, 200 V
Inductor—600 mH, 500 mA
Four AC Ammeters—500 mA
Two AC Voltmeters—250 V

Procedure

Note that because a 200 V supply is used, extra care should be taken to avoid shock.

1   Set each resistance box (R_1, R_2, and R_3) to 500 Ω, then *with the supply remaining unconnected* construct the circuit as illustrated in Fig. L25-1.

2   Connect the 200 V three-phase supply and switch *on*.

3   Record the levels of the line and phase voltages, V_P and V_L. Also record the currents: I_A, I_B, I_C, and I_N.

4   Switch *off* and disconnect the three-phase supply.

5   Adjust the resistors to $R_1 = 500\ \Omega$, $R_2 = 250\ \Omega$, and $R_3 = 750\ \Omega$.

6   Reconnect the supply and switch *on*. Record the new levels of the line and neutral currents.

7   Switch *off*, and disconnect the supply.

8   Reset each resistor to 500 Ω, then modify the circuit to include capacitor C_2 and inductor L_3, as illustrated in Fig. L25-2.

9   Reconnect the supply and switch *on*. Record all current levels once again.

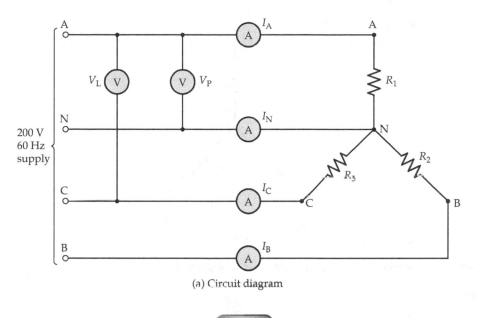

(a) Circuit diagram

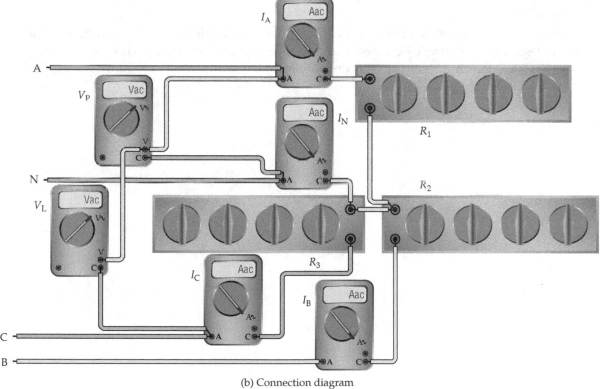

(b) Connection diagram

Figure L25-1   Investigation of Y-connected balanced and unbalanced resistive loads.

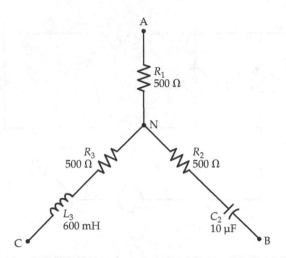

Figure L25-2 Investigation of Y-connected unbalanced impedance loads.

Analysis

Analyze each of the three circuits tested to determine the load and neutral currents for each case. Compare the calculated and measured current levels, and discuss any differences. (Note that the tested circuits are similar to those analyzed in Examples 26-1, 26-2, and 26-3 in the text book.)

Record Sheet L25

Record Sheet   Date _____

Lab. # 25   Three-Phase Y-connected Loads

Procedure 3 Balanced resistive loads

V_p	V_L	I_A	I_B	I_C	I_N

Procedure 6 Unbalanced resistive loads

I_A	I_B	I_C	I_N

Procedure 9 Unbalanced impedance loads

I_A	I_B	I_C	I_N

LABORATORY INVESTIGATION 26
Three-Phase Δ-Connected Loads

Introduction

Delta-connected loads with a three-phase supply are investigated. The loads are (a) balanced resistive, (b) unbalanced resistive, (c) unbalanced impedance. In each case, the line and phase currents are measured. The circuits are analyzed, and the calculated and measured results compared.

Equipment

Three Decade Resistance Boxes—1000 Ω, 1 A
Capacitor—10 μF, 200 V
Inductor—400 mH, 1 A
Six AC Ammeters—1 A

Procedure

Note that because a 200 V supply is used, extra care should be taken to avoid shock.

1 Set each resistance box (R_1, R_2, and R_3) to 500 Ω, then, *with the supply remaining unconnected*, construct the circuit as illustrated in Fig. L26-1.
2 Connect the 200 V three-phase supply and switch *on*.
3 Record the levels of the load and line currents: I_1, I_2, I_3, I_A, I_B, and I_C.
4 Switch *off* and disconnect the three-phase supply.
5 Adjust the resistors to R_1 = 500 Ω, R_2 = 1000 Ω, and R_3 = 750 Ω.
6 Reconnect the supply, switch *on*, and record the new line and load currents.
7 Switch *off*, and disconnect the supply.
8 Reset each resistor to 500 Ω, then modify the circuit to include capacitor C_3 and inductor L_1, as illustrated in Fig. L26-2.
9 Reconnect the supply and switch *on*, and record all current levels once again.

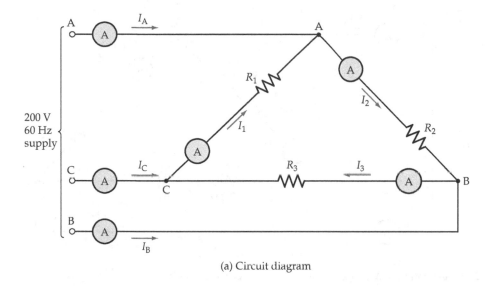

(a) Circuit diagram

(b) Connection diagram

Figure L26-1 Investigation of Δ-connected balanced and unbalanced resistive loads.

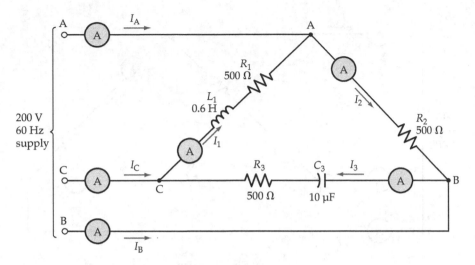

Figure L26-2 Investigation of Δ-connected unbalanced impedance loads.

Analysis

Analyze each of the three circuits tested to determine the line and load currents. Compare the calculated and measured current levels, and discuss any differences. (Note that the tested circuits are similar to those analyzed in Examples 26-4, 26-5, and 26-6 in the text book.)

Record Sheet L26

Record Sheet   Date _____

Lab. # 26   Three-Phase Δ-connected Loads

Procedure 3 Balanced resistive loads

I_1	I_2	I_3	I_A	I_B	I_C

Procedure 6 Unbalanced resistive loads

I_1	I_2	I_3	I_A	I_B	I_C

Procedure 9 Unbalanced impedance loads

I_1	I_2	I_3	I_A	I_B	I_C

LABORATORY INVESTIGATION 27
Phase Sequence Testing

Introduction

A three-phase supply is applied to a Y-connected load consisting of two lamps and an *RL* branch. The line currents are measured, and the relative brightness of the lamps is noted. The phase sequence of the supply is reversed, the line currents are again measured, and the relative brightness of the lamps is noted once again. The circuit is analyzed for each case, and the calculated and measured results are compared.

Equipment

Inductor—600 mH, 1 A
Three AC Ammeters—1 A
Two 60 W, 115 V Lamps
Resistor—100 Ω, 100 W

Procedure

Note that because a 200 V supply is used, extra care should be taken to avoid shock.

1 *With the supply still unconnected*, construct the circuit as shown in Fig. L27-1.
2 Connect the three-phase supply and switch *on*.
3 Record the level of the line currents: I_A, I_B, and I_C, and note the relative brightness of lamps A and B.
4 Switch *off* and disconnect the supply.
5 Disconnect the line cables to terminals B and C. Interchange these cables and reconnect them to the terminals. The cable previously connected to terminal C should now be connected to terminal B, and vice versa.
6 Reconnect the three-phase supply and switch *on*.
7 Record the new levels of the line currents: I_A, I_B, and I_C. Again note the relative brightness of lamps A and B.

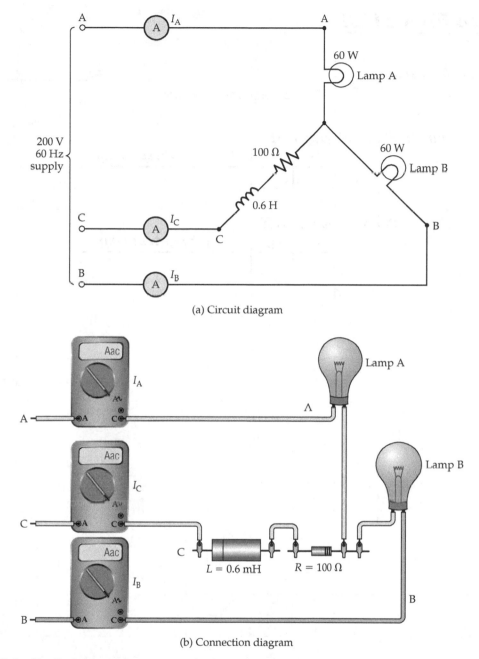

(a) Circuit diagram

(b) Connection diagram

Figure L27-1 Circuit of phase sequence tester for three-phase supplies.

Analysis

Analyze the circuit to determine the load currents for each phase sequence. Compare the calculated and measured currents. (Note that the circuit investigated is similar to the circuit in Examples 26-9 and 26-10 in the text book.)

Record Sheet L27

Record Sheet   Date _____

Lab. # 27   Phase Sequence Testing

Procedure 3   Phase sequence ABC

I_A	I_B	I_C	Brightest lamp

Procedure 4   Phase sequence ACB

I_A	I_B	I_C	Brightest lamp

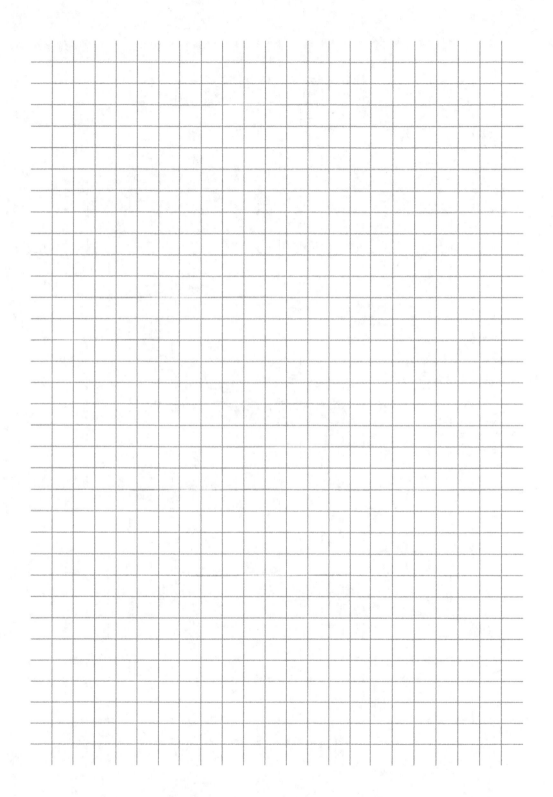

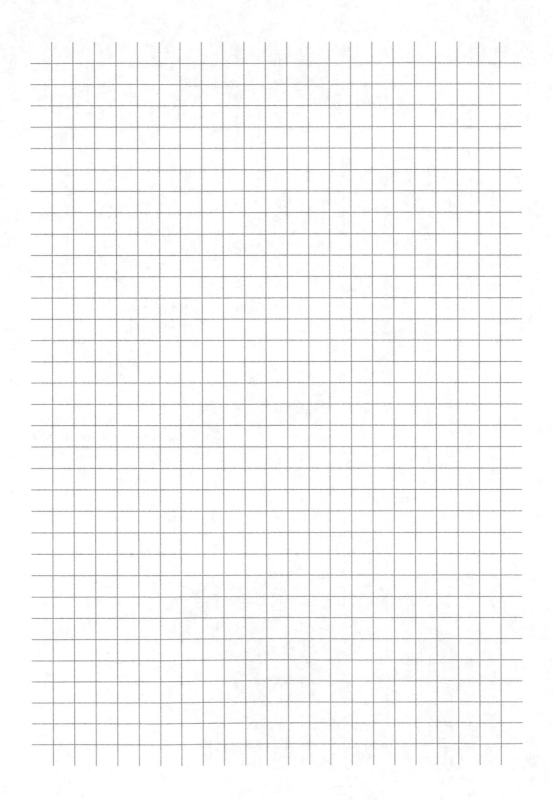

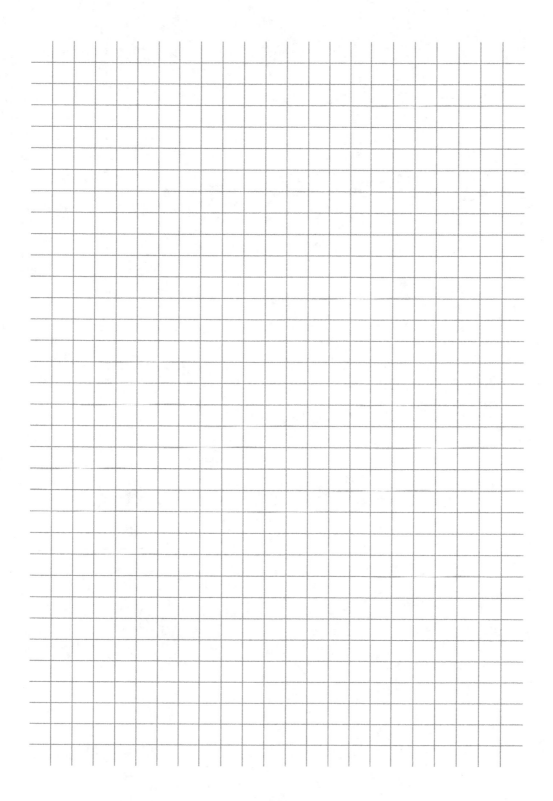

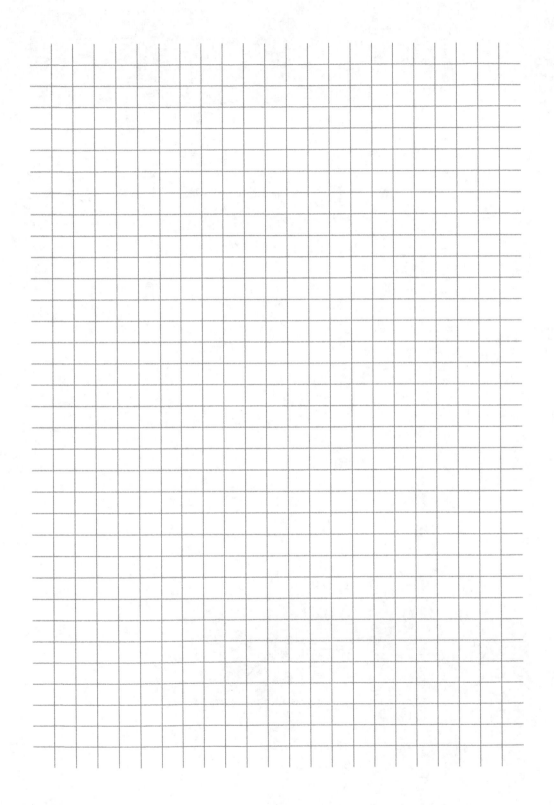

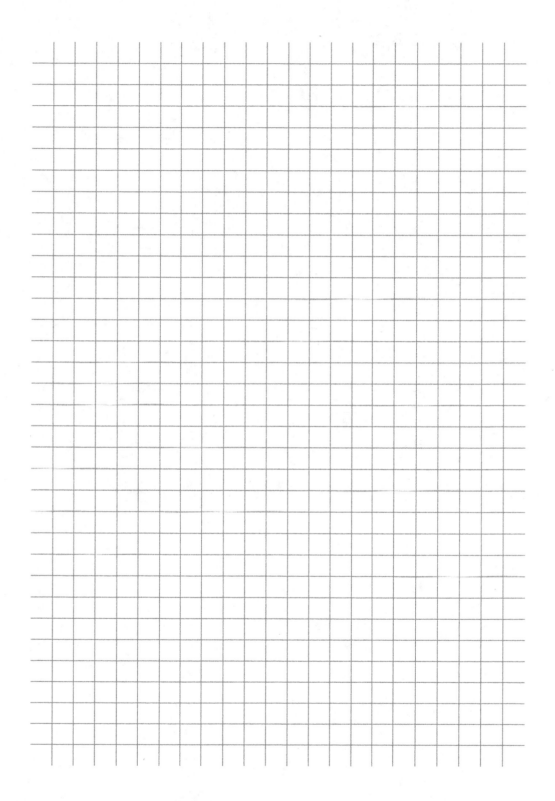

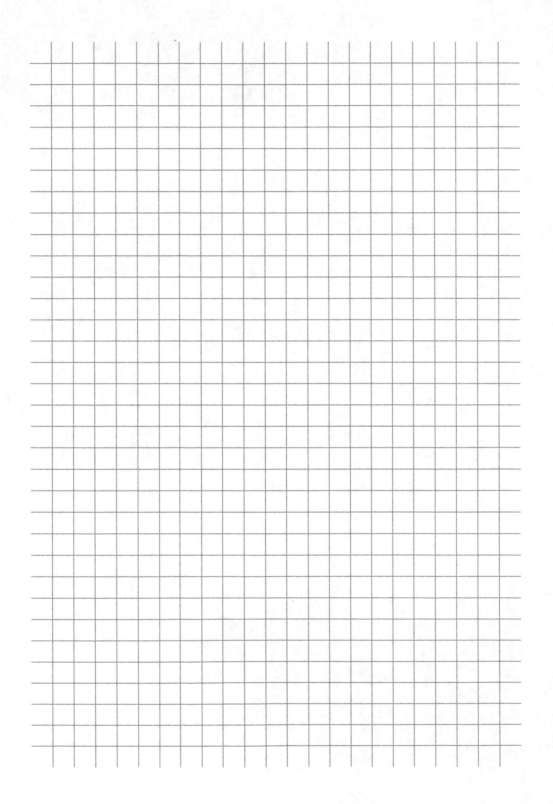